An Introduction to
Mössbauer Spectroscopy

Contributors

Jacques Danon
Professor, Centro Brasileiro de Pesquisas Fisicas, Rio de Janeiro, Brasil

Peter G. Debrunner
Associate Professor of Physics, Department of Physics, University of Illinois, Urbana, Illinois 61801

Hans Frauenfelder
Professor of Physics, Department of Physics, University of Illinois, Urbana, Illinois 61801

Ulrich Gonser
Professor, Institut für Metallphysik und Metallkunde, Universität des Saarlandes, 66 Saarbrücken 15, Germany

David W. Hafemeister
Associate Professor of Physics, Department of Physics, California State Polytechnic College, San Luis Obispo, California 93401

Rolfe H. Herber
Professor of Chemistry, School of Chemistry, Rutgers University, The State University of New Jersey, New Brunswick, New Jersey 08903

Robert L. Ingalls
Associate Professor of Physics, Department of Physics, University of Washington, Seattle, Washington 98105

Leopold May
Associate Professor of Chemistry, Department of Chemistry, The Catholic University of America, Washington, D.C. 20017

Jon J. Spijkerman
Research Physicist, Analytical Chemistry Division, National Bureau of Standards, Washington, D.C. 20234

John C. Travis
Research Physicist, Analytical Chemistry Division, National Bureau of Standards, Washington, D.C. 20234

An Introduction to Mössbauer Spectroscopy

Edited by

Leopold May

Department of Chemistry
The Catholic University of America
Washington, D.C.

PLENUM PRESS • NEW YORK—LONDON • 1971

Library of Congress Catalog Card Number 76-137011

ISBN 978-1-4684-8913-2   ISBN 978-1-4684-8911-8 (eBook)
DOI 10.1007/978-1-4684-8911-8

© 1971 Plenum Press, New York
Softcover reprint of the hardcover 1st edition 1971
A Division of Plenum Publishing Corporation
227 West 17th Street, New York, N.Y. 10011

United Kingdom edition published by Plenum Press, London
A Division of Plenum Publishing Company, Ltd.
Donington House, 30 Norfolk Street, London W.C. 2, England

Preface

The initial impetus for this text occurred when we were searching for a single book that could be recommended to the attendees at the Mössbauer Spectroscopy Institute at The Catholic University of America. This Institute is an introductory course on the theory and interpretation of Mössbauer spectroscopy for workers in industrial, academic, and government laboratories. None of the books available adequately covered the breadth and scope of the lectures in the Institute. A list of these books and review articles is included in Appendix C. To meet our needs, we undertook the creation of this text.

The chapters are based upon the lectures given at the various Institutes from 1967 to 1969. Most of the lectures were recorded and transcripts sent to the lecturers, who then prepared the manuscripts, using the transcripts as a guide so as to retain the style developed during the lecture. Each chapter is written in the style of the authors. As the editor, my main task was to maintain uniformity of format and nomenclature. A list of nomenclature used in this volume is reproduced in Appendix A. We hope that this list will be used particularly by new investigators and teachers of Mössbauer spectroscopy so that future literature will employ a uniform system.

The text is written primarily to introduce scientists, both at the graduate student level and in active research, to areas in which Mössbauer spectroscopy may be of assistance. Each author is an active research worker, and many are regarded as leaders in their particular aspect of the subject matter. They bring to their writing both specific knowledge of their research and the desire to transmit adequately the results and principles to the reader.

I wish to thank the authors for their efforts and suggestions in making this endeavor possible and also the secretarial staff at the Chemistry Department of Catholic University for their assistance, particularly Mrs. Olivia Messer and Miss Judy Williams.

Leopold May

Contents

Chapter 1

Introduction to the Mössbauer Effect

Peter G. Debrunner and Hans Frauenfelder

University of Illinois
Urbana, Illinois

At the present time there already exist a number of good introductions to the Mössbauer effect and its applications. There is, therefore, no need to write another, similar, paper, which would in large measure be a copy of earlier ones. In the present chapter we try to avoid this duplication by stressing the physical background and by giving the simplest possible pictures. These pictures may help in explaining the Mössbauer effect to nonphysicists, and they may even amuse some physicists.

1. PICTORIAL DESCRIPTION

We begin by giving an easy-to-understand classical example that helps

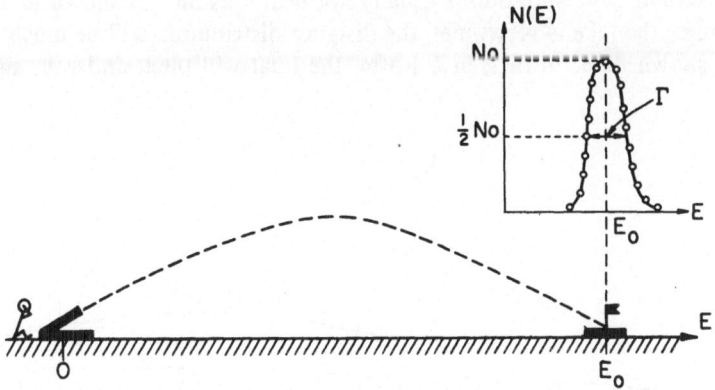

Figure 1.   Straggling in the distance of projectiles shot by a fixed cannon. $N(E)$ is the number of projectiles observed at the distance E.

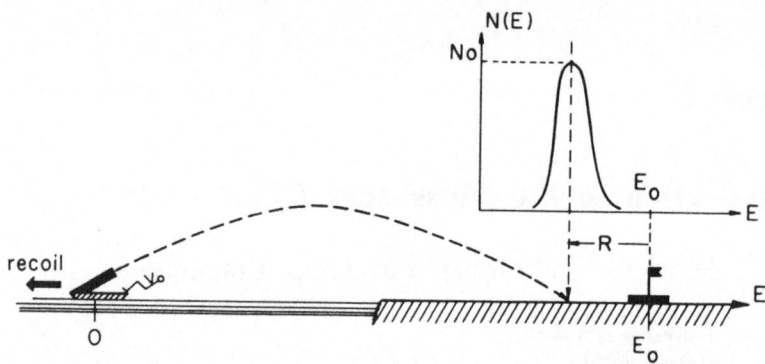

Figure 2. If the cannon is not fixed, it can recoil. As a result, projectiles fall short of target by a distance R.

to explain many of the features of the Mössbauer effect. We consider a cannon, fixed rigidly to the earth, and a target at a distance E_0. If we trace a large number of shots and measure their distances E, we find that not all have flown the same distance; there exists a straggling as sketched in Figure 1. We characterize the distribution by the full width at half-height and we call this quantity Γ, the "natural line width."

Next we mount our cannon on a small boat and try to hit the same target again. We expect the same natural line width, and indeed it appears. However, the small boat recoils in order to conserve momentum and the shots fall short, as indicated in Figure 2. The distance they fall short, R, can be calculated from energy and momentum conservation.

Actually, the situation is usually not nearly as nice as shown in Figure 2. Unless the lake is very quiet, the distance distribution will be much wider than shown here. In fact, in a storm, the boat will pitch and roll, and the

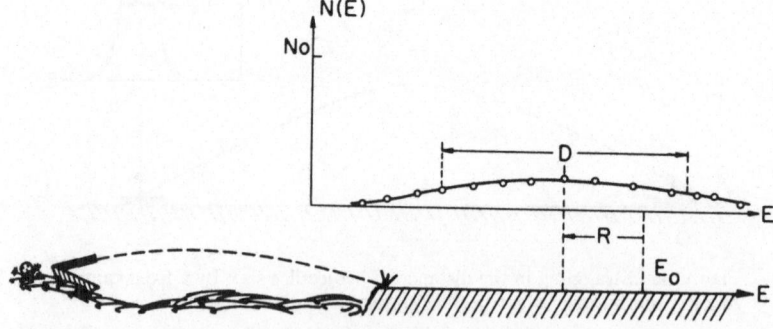

Figure 3. Shots fired from a cannon mounted on a small boat in a rough sea experience both a recoil (R) and a Doppler broadening (D).

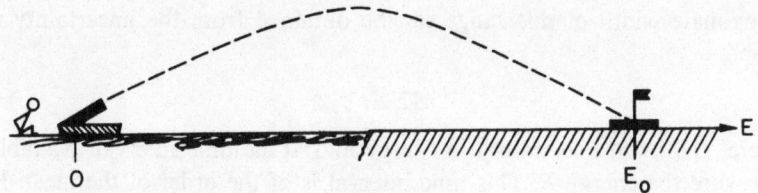

Figure 4. If the lake is frozen, Doppler broadening and recoil are avoided, and aiming is simple. Straggling still persists.

distribution will experience an additional broadening, which we call the "Doppler broadening," shown in Figure 3.

Is there a way to avoid the recoil R and the Doppler broadening D? Rudolf Mössbauer found the answer in a somewhat different context. If we wait till the lake is frozen, Doppler broadening and recoil will disappear! (See Figure 4.)

2. BACKGROUND CONCEPTS

The explanation of the Mössbauer effect in nuclear transitions follows our simple analog reasonably closely. Before explaining the effect, we describe some essential concepts.

2.1. Natural Line Width

From many experiments in nuclear and atomic physics we are used to the existence of energy levels. We normally indicate these levels as lines and assume that the energy of a state is given by E_0, the solution of the Schrödinger equation for a particular problem. Actually, however, these energy levels have a certain width, as indicated in Figure 5. The energy E of the level then is not "sharp," but is spread over a certain energy range. The ap-

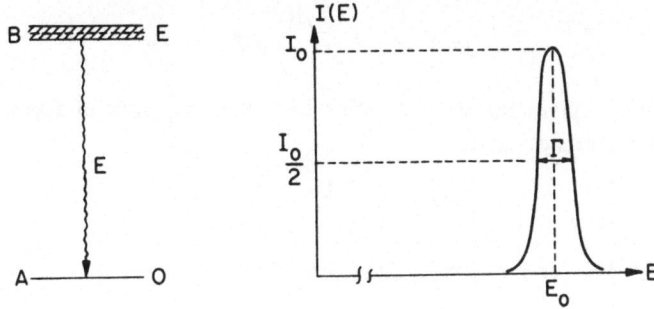

Figure 5. Natural line width. The energy of the level with a mean life τ has a width $\Gamma = \hbar/\tau$.

proximate width of this range can be obtained from the uncertainty relation.

$$\Delta E \Delta t \geq \hbar \tag{1}$$

Here, ΔE is the uncertainty in energy and Δt the time interval available to measure the energy E. This time interval is of the order of the mean life τ of the state under consideration; if our experiment takes much longer than τ, the state has disappeared! Setting $\Delta t \approx \tau$, we get for the approximate width $\Gamma = \Delta E$ of the level

$$\Gamma = \hbar/\tau \tag{2}$$

Many years ago Weisskopf and Wigner [1] treated this problem correctly and found that Eq. (2) indeed holds if Γ denotes the full width of the energy distribution at half-height. More exactly they found that the line has a Lorentzian or Breit–Wigner shape and can be described by

$$I(E) = \text{const} \, \frac{\Gamma}{2\pi} \, \frac{1}{(E - E_0)^2 + (\Gamma/2)^2} \tag{3}$$

The form of the line shape can be understood classically. A state decaying with a mean life $\tau = \hbar/\Gamma$ classically corresponds to an exponentially decaying wave train with amplitude

$$\psi(t) = \psi_0 e^{-i\omega_0 t} e^{-\Gamma t/2\hbar}, \quad \text{for } t \geq 0 \tag{4}$$
$$0 \qquad , \quad \text{for } t < 0$$

where ω_0 is the average frequency. The intensity of a wave is proportional to the absolute square of its amplitude; Eq. (4) shows that $I(t)$ is given by

$$I(t) \propto |\psi(t)|^2 = |\psi_0|^2 e^{-\Gamma t/\hbar} \tag{5}$$

and indeed decays with a mean life τ. However, the wave is not monochromatic; to obtain the frequency distribution, we expand $\psi(t)$ as

$$\psi(t) = \frac{1}{\sqrt{2\pi}} \int_{-\infty}^{+\infty} d\omega \Phi(\omega) e^{-i\omega t} \tag{6}$$

Here, $\Phi(\omega)$ gives the weight with which the frequency ω appears in $\psi(t)$. Fourier inversion gives

$$\Phi(\omega) = \frac{1}{\sqrt{2\pi}} \int_{-\infty}^{+\infty} dt \psi(t) e^{i\omega t} \tag{7}$$

or, with Eq. (4) and after integration

$$\Phi(\omega) = \frac{\psi_0}{\sqrt{2\pi}} \, \frac{1}{i(\omega_0 - \omega) + \Gamma/2\hbar} \tag{8}$$

The intensity distribution

$$I(\omega) \propto |\Phi(\omega)|^2 = |\psi_0|^2 \frac{1}{2\pi} \frac{1}{(\omega_0 - \omega)^2 + (\Gamma/2\hbar)^2} \tag{9}$$

coincides with the quantum mechanical expression given in Eq. (3).

For a stationary state, the mean life is infinite, and Eq. (3) shows that such a state is indeed "sharp." The photons emitted in a transition from an excited state with mean life τ to a stationary ground state then have an energy distribution that is also given by Eq. (3); the distribution is sketched in Figure 5. In our analogy, the line width then corresponds to the straggling shown in Figure 1. There is one important difference, however. Cannons have a rather large "natural line width," but most nuclear and atomic transitions have a very small one. For an excited state with mean life τ, we get from Eq. (2)

$$\Gamma(\text{in eV}) = 6.58 \times 10^{-16}/\tau(\text{sec}) \tag{10}$$

For a mean life of 10^{-8} sec, we obtain a natural line width of 6.58×10^{-8} eV. If the transition energy is 66 keV, the ratio of line width to energy, often denoted with $1/Q$, is very small

$$1/Q = \Gamma/E_0 = 10^{-12}$$

2.2. Recoil Energy Loss

The photons emitted in the transition shown in Figure 5 possess the mean energy E_0 if they are emitted from a system of infinite mass. In general, however, there will be a recoil energy loss R just as indicated in the classical case of Figure 2. To calculate R, we assume that the photon is emitted by a nucleus of mass M that is at rest before the decay (Figure 6). Momentum conservation then gives

$$\mathbf{p}_{\text{nucleus}} = -\mathbf{p}_{\text{photon}} \tag{11}$$

The magnitude of the photon momentum is connected with the photon energy by

$$p_{\text{photon}} = E_{\text{photon}}/c \tag{12}$$

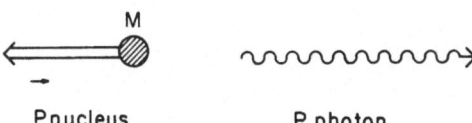

Figure 6. Recoil energy loss caused by the recoil of the decaying nucleus.

where c is the velocity of light. Since nuclei are very heavy (when their rest energy is compared to the decay energy E_0), we can use the nonrelativistic approximation to connect the magnitude of the momentum with the recoil energy R

$$R = p^2_{nucleus}/2M \qquad (13)$$

Furthermore, because R will be small compared to E_0, we can set $E_{photon} = E_0$ in Eq. (12), and then from Eqs. (11) to (13)

$$R = E_0^2/2Mc^2 \qquad (14)$$

For a given decay energy E_0 and a given nuclear mass M, R can be calculated easily from Eq. (14). For a quick evaluation, we rewrite Eq. (14) into the numerical form

$$R(\text{in eV}) = 5.37 \times 10^{-4} E_0^2(\text{in keV})/A \qquad (15)$$

where A is the atomic number of the decaying nucleus.

As a numerical example, we use a nucleus with $A=100$, with an energy $E_0=66$ keV, and with a mean life $\tau=10^{-8}$ sec. Equation (15) then gives $R=0.02$ eV. The recoil energy loss is indeed small compared to the decay energy. However, it is very large compared to the natural line width of our example (6.6×10^{-8} eV). In terms of our Figure 2 we can say that the projectile falls far short of its target! We will see in the next section why this "falling short" is so important.

2.3. Resonance and Resonance Fluorescence

To discuss resonance fluorescence, we consider first another classical problem, a resonator. One of the best known resonators is the swing (Figure 7). We all know from experience that maximum excursion is obtained if the driving force is in resonance with the proper frequency of the swing. The excitation of an atomic or nuclear level also shows such a behavior. The *maximum* absorption cross section occurs if an incident gamma

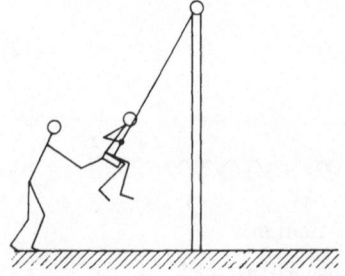

Figure 7.   A well-known resonator.

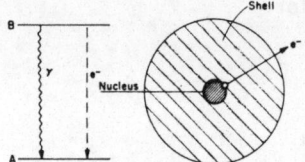

Figure 8. Gamma emission and internal conversion compete in most transitions that are useful for Mössbauer work.

ray has the energy E_0, the excitation energy of the atomic or nuclear level; it is given by

$$\sigma_0 = 2\pi\lambda^2 \frac{2J_B + 1}{2J_A + 1} \frac{\Gamma_\gamma}{\Gamma} \tag{16}$$

where J_A is the spin of the ground state A, J_B the spin of the excited state B, and $2\pi\lambda$ is the wavelength of the incident photon.

The factor Γ_γ/Γ in Eq. (16) requires some additional comments. Frequently, a nuclear level decays not only by gamma emission but also by competing processes. The process that is most important for Mössbauer spectroscopy is internal conversion, and we will restrict our discussion to it. In internal conversion, the excitation energy of the nucleus is transferred directly to the electron shell, and an electron is ejected. Symbolically, the two competing processes—gamma emission and internal conversion—are shown in Figure 8.

The decay properties of level B are no longer sufficiently described by the mean life τ or the width $\Gamma = \hbar/\tau$. We introduce partial lifetimes τ_γ and τ_e and partial line widths $\Gamma_\gamma = \hbar/\tau_\gamma$ and $\Gamma_e = \hbar/\tau_e$. The expression

$$\Gamma_e/\Gamma_\gamma = \tau_\gamma/\tau_e = \alpha \tag{17}$$

gives the ratio of nuclei decaying by internal conversion to those decaying by gamma emission; α is called the conversion coefficient. It is important to realize that the actual line width is the sum of the partial line widths

$$\Gamma = \Gamma_\gamma + \Gamma_e \tag{18}$$

The observed mean life is given by

$$\tau = \hbar/(\Gamma_\gamma + \Gamma_e) \tag{19}$$

and it is this mean life with which photons *and* conversion electrons are emitted. (We can make a simple analog. If a can containing water is emptied through two holes, one large and one small, water will flow through both holes as long as there is some left. The large hole will not dry up faster than the small one.)

The ratio Γ_γ/Γ in Eq. (16) can now be understood: if we try to excite the level B in Figure 8 by shining photons of the correct energy onto the

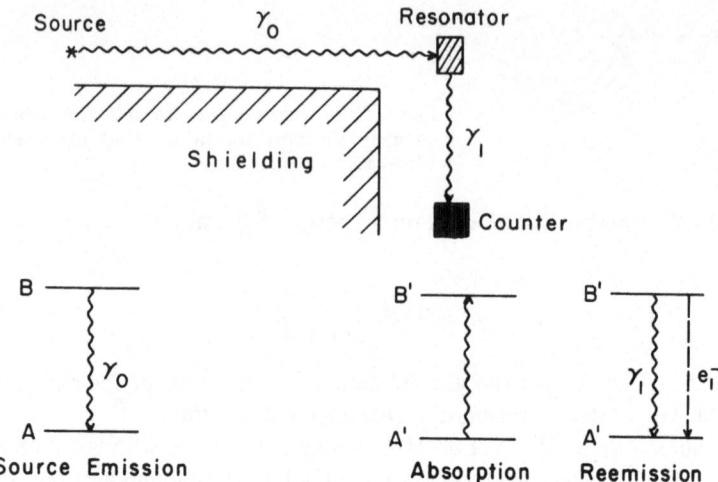

Figure 9.   Nuclear resonance fluorescence experiment.

nucleus, the cross section for excitation by photons will be reduced if the main decay of the level occurs through internal conversion.

To calculate the cross section σ_0, we note that

$$\Gamma_\gamma/\Gamma = 1/(1 + \alpha) \tag{20}$$

With Eq. (12) and $\lambdabar = \hbar c/E$, we then rewrite Eq. (16) as

$$\sigma_0(\text{in b}) = \frac{2.45 \times 10^9}{E^2(\text{in keV})} \frac{2J_B + 1}{2J_A + 1} \frac{1}{1 + \alpha} \tag{21}$$

One b is equal to 10^{-24} cm², and the cross section σ_0 can therefore become extremely large.

The large cross section suggests a resonance experiment as sketched in Figure 9: The photons from a radioactive source are absorbed by a resonator. If the nuclei in the resonator are identical to the source nuclei, we expect that the photons are resonantly absorbed and that they excite the resonator nuclei into the level B. Once excited, the nuclei will decay again with the reemission of gamma rays of energy E_0. The reemission can then be observed from the side. The complete process is called "nuclear resonance fluorescence." Consider the case where source and resonator are free nuclei at rest. We can think at first that such a situation exists in gases. We will, however, show in the next subsection that gases introduce a new problem; free source and resonator nuclei at rest exist only in thought experiments. Even with free nuclei at rest we are in difficulties. The incident gamma ray γ_0 only has an energy $E_0 - R$. Moreover, if a nucleus of mass M and initially at rest

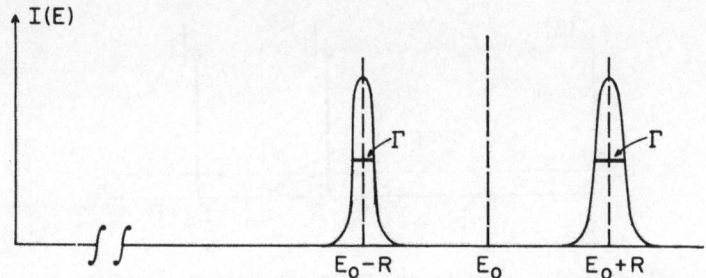

Figure 10. E_0 denotes the energy of the level B. The photon emitted from a free nucleus at rest only possesses the energy E_0-R; in order to excite the level B in the resonator it requires the energy E_0+R.

absorbs a photon of momentum p, it will recoil with this momentum p. The recoil energy $R=p^2/2M$ then is not available to excite the nucleus. In order to excite a level of energy E_0, the incident gamma ray must have an energy E_0+R. The situation is thus as shown in Figure 10. Because for all cases of interest the recoil energy R is much larger than the natural line width Γ, no excitation can take place. Before we discuss how the Mössbauer effect remedies this situation, we treat the actual case of gamma rays emitted by gases or solids.

2.4. Doppler Broadening

In a gas or a solid, nuclei are not at rest; they move with rather large velocities. In a gas, the velocity can be calculated easily from classical considerations: The kinetic energy of an atom (or molecule) is given by

$$E_{\text{kin}} = (1/2)Mv_0^2 = (3/2)kT \tag{22}$$

Here, T is the temperature (^{0}K) and k is the Boltzmann constant (k=8.62 $\times 10^{-5}$ eV/^{0}K). At room temperature, typical velocities are of the order of a few hundred m/sec. Such a velocity leads to a *Doppler broadening*. The energy of a gamma ray emitted by a source moving with velocity component v_r along the direction of emission is shifted by an amount ΔE given by

$$\Delta E = \frac{v_r}{c} E_0 \tag{23}$$

In a gaseous source the velocity of the emitting atom will be directed at random with respect to the direction source resonator in Figure 9. The velocity along the direction of emission will therefore vary from $+v_0$ to $-v_0$, and the line shape observed from a large number of decaying atoms will be broadened by an amount

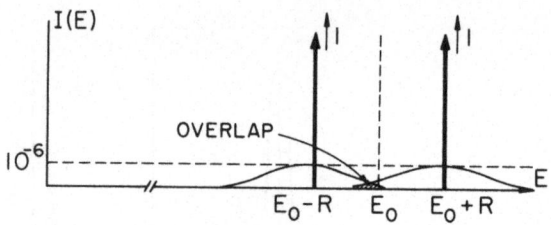

Figure 11. Emission and absorption lines are Doppler broadened, and a small overlap exists. The Doppler broadened lines are much less peaked than the natural ones.

$$\bar{D} \cong 2\frac{v_0}{c}E_0 \tag{24}$$

For our standard example ($E_0=66$ keV, $A=100$, $T=300°K$) this broadening is about 0.1 eV, or of the same order as R. Figure 10 should therefore be replaced by Figure 11. Emission and absorption lines now overlap somewhat and a small amount of resonance fluorescence is expected [2]. Unfortunately, we have paid a high price for the overlap. The line height is reduced by the factor Γ/D, which in our example is about 10^{-6}. This reduction drastically affects the number of observed fluorescence photons. One way to see this reduction is by using Eq. (16). We have not changed Γ_γ, the partial gamma-ray width. The total width Γ, however, has increased by about a factor of 10^6, and the maximum cross section therefore has decreased by the same factor! How can we have the cake and eat it too? Is it possible to observe nuclear resonance fluorescence with nonbroadened lines? Of course, with hindsight and with the help of Figure 4 an answer is easy: we only have to freeze source and resonator properly in order to get sharp and unshifted lines! Indeed, the Mössbauer effect is just such a freezing process, and Rudolf Mössbauer discovered it during his doctoral thesis research. The problem that had been given him by his thesis advisor, H. Maier-Leibnitz, was the study of nuclear resonance fluorescence at low temperatures! However, the concept of freezing here is somewhat more complicated than the freezing of boats on a lake and we first discuss some aspects of the theory of solids before we explain the recoilless emission of gamma rays.

2.5. Einstein Solids

Classically, the atoms of a solid have a kinetic energy essentially in agreement with Eq. (22), and they vibrate around their equilibrium positions with an energy that varies continuously with temperature. However, it was first pointed out by Einstein [3] that a solid is also a quantum mechanical system and that its energy should also be quantized. Einstein assumed for

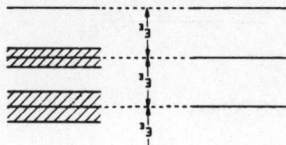

Figure 12.   Energy levels of an Einstein solid. At left are the levels without interaction among the atoms. The levels at right correspond more closely to an actual situation; the excited states are broadened.

simplicity that the energy levels of a solid should be equidistant; the energy spectrum then shows only values

$$E_g + nE_E, \qquad n = 0, 1, \ldots$$

as indicated in Figure 12. (Actually, even in such an Einstein solid the excited levels are broadened because of interactions among the atoms.) The energy levels of a real solid are more complicated than the ones used by Einstein; this fact was pointed out simultaneously by Debye [4] and by Born and von Karman [5]. For our discussion, the Einstein solid is sufficient.

In order to arrive at a crude value for the Einstein energy E_E, we consider a solid, as indicated in Figure 13, and use the famous equation

$$E = \hbar\omega = \hbar v/\lambdabar \tag{25}$$

Here, v is the velocity of the waves in the solid, and $2\pi\lambdabar$ is the wavelength. Figure 13 shows that a reasonable assumption for the wavelength is $\lambda = 2a$, were a is the lattice constant. For v, we take the velocity of sound. We then get

$$E_E = \pi\hbar v/a \tag{26}$$

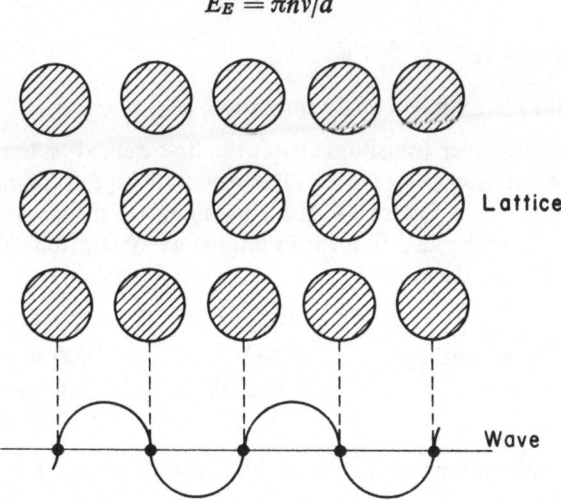

Figure 13.   Propagation of sound through an Einstein solid.

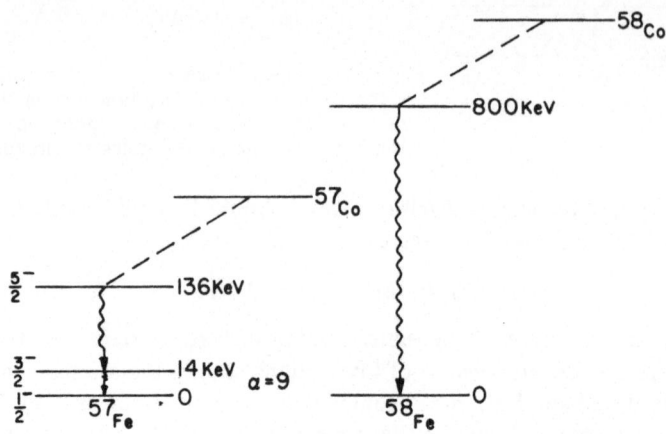

Figure 14.   Decay schemes for ^{57}Fe and ^{58}Fe. Only the essential aspects are shown.

Take iron as an example. It has a sound velocity of 5960 m/sec and a lattice constant of 2.9Å. Equation.(26) then gives

$$E_E(\text{iron}) = 0.04 \text{ eV} \tag{27}$$

Sometimes, the Einstein temperature is given instead of the Einstein energy; it is defined by

$$\theta_E = E_E/k \tag{28}$$

For iron, we get $\theta_E(\text{Fe}) = 500°\text{K}$.

2.6.   Recoil-Free Emission of Gamma Rays

We now consider transitions from the first excited to the ground state in two different isotopes of iron. The essential aspects of the two decay schemes are given in Figure 14. If we assume that the decaying ^{57}Co and ^{58}Co nuclei are embedded in an iron lattice, we obtain the following parameters for the two decays.

	^{57}Fe	^{58}Fe
Decay energy	14 keV	800 keV
R	0.002 eV	6 eV
E_E	0.04 eV	0.04 eV

From these values, we note a crucial difference between ^{57}Fe and ^{58}Fe: For the low-energy transition, the recoil energy is much smaller than the Einstein energy, whereas the opposite is true for the 800-keV photons. This situation is depicted in Figure 15.

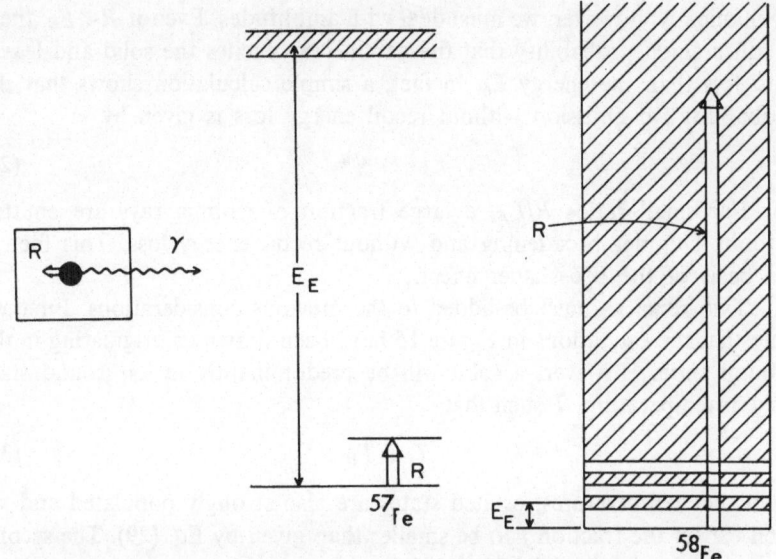

Figure 15. Emission of gamma rays from nuclei embedded in an iron lattice. For ^{57}Fe, the recoil energy R is small compared to the Einstein energy. For ^{58}Fe, R excites the lattice to such a high energy that the individual energy levels overlap and the lattice behaves like a classical solid.

We first describe the situation "semiclassically." We assume that the solid is a quantum mechanical system with levels as shown in Figure 15, but that the emission follows classical laws. The 14-keV transition of ^{57}Fe then is emitted without recoil: The minimum amount of energy that the solid can accept is E_E. Since R is much smaller than E_E, the solid cannot accept this amount R. The gamma ray therefore escapes with the full energy E_0. Moreover it is not Doppler broadened. Doppler broadening comes from thermal excitation of the solid; the 14-keV gamma ray escapes without changing the internal energy of the solid and hence has the natural line shape. In terms of our analog, Figure 1, we have "frozen" the emitting nucleus into the solid. The situation is different for the 800-keV transition. Here R is very large compared to E_E, and the solid is excited very highly. At such high excitation energies, the discrete Einstein levels overlap, and the energy spectrum is a continuum. The solid therefore can accept any particular recoil energy R, and the iron nucleus can emit its photon as if it were free. The emitted photon therefore suffers the expected energy loss. Moreover, classically, the atoms vibrate in the solid, and this vibration gives rise to a Doppler broadening of the photon line.

Classically, there are only two possibilities: if $R < E_E$, the gamma ray escapes without energy loss; if $R \geq E_E$, recoil energy loss can occur. Quantum mechanically, however, we must deal with amplitudes. Even if $R \ll E_E$, there is still a small probability that the gamma ray excites the solid and leaves with less than the energy E_0. In fact, a simple calculation shows that the probability for emission without recoil energy loss is given by

$$f = e^{-R/E_E} \tag{29}$$

For small ratios R/E_E, a large fraction of gamma rays are emitted without Doppler broadening and without recoil energy loss. This fact is the basis of the Mössbauer effect.

Two remarks must be added to the previous considerations. First we note that the excitations in Figure 15 have been drawn as originating in the ground state. However, a solid will be predominantly in its ground state only at temperatures T such that

$$T \ll T_E \tag{30}$$

At higher temperatures, excited states are also strongly populated and we then expect the fraction f to be smaller than given by Eq. (29). The second remark refers to the Einstein model. Although this model fits some data well, actual lattice spectra are more complicated. We will discuss both these aspects in more detail in the section on theory.

3. THE MÖSSBAUER EFFECT

With the background material given in Sections 1 and 2, it is easy to understand the main features of the Mössbauer effect. As before, we first give a naive analog and then sketch how actual experiments are performed.

3.1. Pictorial Approach

We ask the next question again with reference to Figure 4. With the frozen boat we got rid of recoil energy loss and Doppler broadening. How can we study the "natural line shape" of the cannon on the frozen boat? We assume that we get a signal if the target is hit and that the signal is proportional to how closely the target has been hit. We then simply move the cannon on the frozen boat forward and backward in small steps and measure the signal as a function of the displacement of the cannon. The result

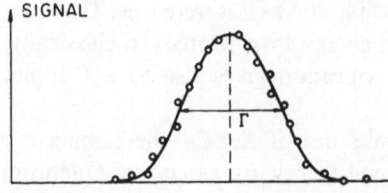

Figure 16. Signal as a function of displacement of the cannon shown in Figure 4.

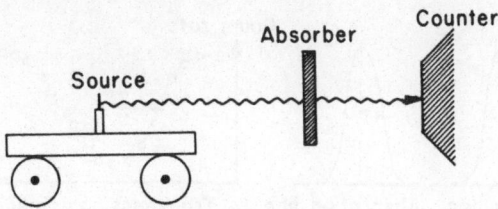

Figure 17. Arrangement to observe the Mössbauer effect.

will look as in Figure 16. The width of the signal gives the "natural line shape" of the cannon. Of course, the maximum of the curve will only be at zero if the cannon was already positioned in the optimum place. In general, there will be a shift between the zero position and the maximum of the curve.

3.2. Observation of the Mössbauer Effect

The Mössbauer effect can be understood in terms of our previous analogy and with reference to Figure 17. In this figure, the source emits gamma rays from a certain nuclide, say ^{57}Fe. A fraction f of these gamma rays is emitted without recoil energy loss and without Doppler broadening. After passing through an absorber containing ^{57}Fe, the 14-keV photons are counted in a detector. The fraction $(1 - f)$ of the 14-keV photons undergoes normal absorption in the absorber (photoeffect and Compton effect), and the nonabsorbed part is counted. The recoilless gamma rays also suffer this normal absorption. In addition, however, a fraction f' (or a fraction ff' of the total gamma rays incident on the absorber) gives rise to resonant absorption. These photons excite the ^{57}Fe nuclei in the absorber, and they are therefore removed from the beam. (Of course, the nuclei thus excited decay again and reemit either a conversion electron or a 14-keV gamma ray. Reemission occurs in all directions, and the fraction of reemitted gamma rays that is measured in the detector can usually be neglected.) What now is the total effect of the absorption of the recoilless photons? The first effect, and the one that was first noticed by Mössbauer, is the excess absorption. More photons are removed from the beam than is anticipated on the basis of well-established laws of photo- and Compton effect. A second effect, also first observed by Mössbauer, is more dramatic and more important for all applications of the Mössbauer effect. We remarked in Section 3.1. that the "natural line width" can be observed by properly shifting the cannon on the boat. Now distance in our analog corresponds to energy for the real photons. Can we shift the energy of the emitted photons? If we can do it, what happens?

Shifting the energy by a small amount is easy. Equation (23) shows that the energy of a photon emitted by a moving source is displaced by an amount

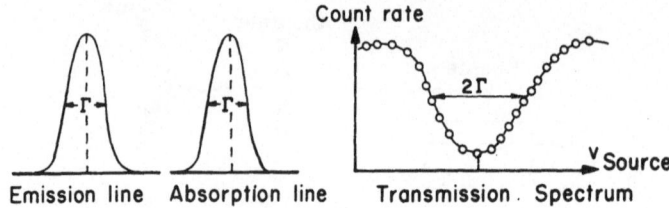

Figure 18. Emission, absorption, and transmission lines in a simple Mössbauer arrangement.

$(v/c)E_0$. If the source is moved with a constant velocity v, the emitted photons will have an energy $E'_0 = E_0(1 + v/c)$.

What does this energy shift do? Emission and absorption lines both show (in the ideal case) a Lorentzian shape with width Γ. If their centers coincide, the absorption will be maximized. If one of the lines is shifted, the overlap will be smaller, the absorption will be smaller, and the counting rate in the detector will increase. It is easy to see that the resulting transmission line is again a Lorentzian, but with width 2Γ. These lines are shown in Figure 18.

The observation of the line shape through observation of the transmitted photon intensity as a function of source velocity is crucial for almost all Mössbauer experiments. Before describing what can be learned from a Mössbauer spectrum, we return to the theory of recoilless emission.

4. THEORY

In the previous sections, we have given a description of the Mössbauer effect in simple terms. The ideas of Sections 1 and 2 are sufficient for an understanding of the physical basis, but they do not permit calculations that can be compared with actual experiments. In the present section, we remedy the situation and provide a firmer foundation. We follow unpublished notes by Lipkin [7] in the derivation of the exact expression for the recoilless fraction f and the discussion of a one-dimensional harmonic oscillator. Finally, we apply our results to the Debye model, a more realistic approximation of a crystal lattice.

We begin by considering the radioactive nucleus, its atom, and the entire solid in which this atom is embedded, as *one* quantum-mechanical system. To treat this complicated system without approximations is difficult, and we therefore assume that the wave function can be written as a product

$$\psi_{\text{total}} = \Phi_{\text{nucleus}}\psi_{\text{solid}} \tag{31}$$

The separation into a product of wave functions is possible because nuclear forces are strong, but have a short range; the nuclear wave function then is

not influenced by the state of the solid, and the wave function of the solid
is not affected by the state of the nucleus. We now face the following problem: Initially, the nucleus is in an excited state, and the solid is in a stationary state, which we denote with ψ_i. The emission of a gamma ray of momentum $\mathbf{p}_0$ leaves the nucleus in its ground state and the solid in a state ψ_f. In
general, ψ_f will be different from ψ_i; the change $\psi_i \rightarrow \psi_f$ is caused by the
recoil that is imparted to the solid during the gamma emission. A certain
probability f exists for emission without change in lattice state. Somewhat
improperly, processes with $\psi_f = \psi_i$ are called "recoilless." We will now derive
an expression for f.

The stationary states of the solid are given by the Schrödinger equation

$$Hu_n = \varepsilon_n u_n \qquad (32)$$

where H is the Hamiltonian of the solid and u_n is a stationary state of the
solid with energy ε_n. We assume that the functions u_n form a complete
orthonormal set. Before the gamma emission, the solid is in a stationary state
with energy ε_i; hence, we have $\psi_i = u_i$. Furthermore, we assume that the
solid is at rest so that u_i is an eigenstate of the momentum operator $-i\hbar\nabla$
with eigenvalue 0

$$Hu_i = \varepsilon_i u_i, \qquad -i\hbar\nabla u_i = 0 \qquad (33)$$

After emission of a photon with momentum $\mathbf{p}_0 = \hbar\mathbf{k}_0$, the solid is in a state
ψ_f. Since the recoil energy even for very energetic gamma rays is too small
to eject an atom from the solid, the entire solid must take up the recoil
momentum $-\mathbf{p}_0$. The state described by ψ_f therefore must be an eigenstate
of the momentum operator, with eigenvalue $-\mathbf{p}_0$

$$-i\hbar\nabla\psi_f = -\mathbf{p}_0\psi_f \qquad (34)$$

To find ψ_f we first expand the wave function of the initial state of the
solid in plane waves[1]

$$u_i = \sum_{\mathbf{k}} c_{\mathbf{k}i} e^{i\mathbf{k}\cdot\mathbf{x}} \qquad (35)$$

The expansion coefficients $c_{\mathbf{k}i}$ can be computed if the wave function ψ_i is
known explicitly. However, we do not need explicit expressions here.

The momentum state of the solid before and after gamma emission differs by the recoil momentum; the recoil changes each momentum in the
expansion Eq. (35) from $\mathbf{k}$ to $\mathbf{k} - \mathbf{k}_0$. The state ψ_f is hence given by

$$\psi_f = \sum_{\mathbf{k}} c_{\mathbf{k}i} e^{i(\mathbf{k} - \mathbf{k}_0)\cdot\mathbf{x}}$$

---

[1] Strictly speaking we should use a Fourier integral here instead of the Fourier sum.
However, the sum is sufficiently general for our purposes and the generalization is left
as a problem.

or with Eq. (35)

$$\psi_f = e^{-i\mathbf{k}_0 \cdot \mathbf{x}} u_i \tag{36}$$

This wave function satisfies the conditions Eq. (34) and represents a solid recoiling with momentum $-\mathbf{k}_0$

We have noted above that the solid, after the gamma emission, is no longer in an eigenstate of H. To compute the probability of finding the solid in a given energy state, we expand ψ_f in terms of the energy eigenfunctions u_n

$$\psi_f = \sum_n c_n u_n \tag{37}$$

The probability of finding the solid in a state with energy ε_r is then given by $|c_r|^2$, or

$$|c_r{}^2| = \left| \int d^3x u_r{}^* \psi_f \right|^2 = \left| \int d^3x u_r{}^* e^{-i\mathbf{k}_0 \cdot \mathbf{x}} u_i \right|^2 \tag{38}$$

where we have used Eq. (36). If we select $r = i$, the solid has the same energy after the gamma emission as before; the photon must have carried away the full transition energy and thus was emitted without recoil-energy loss and without Doppler broadening. The quantity

$$f = |c_i|^2 = \left| \int d^3x u_i{}^* e^{-i\mathbf{k}_0 \cdot \mathbf{x}} u_i \right|^2 \tag{39}$$

is therefore the recoil-free fraction that we wanted to calculate. This expression has a straightforward interpretation. Since

$$\varrho(\mathbf{x}) = u^*(\mathbf{x}) u(\mathbf{x}) \tag{40}$$

is the probability density, Eq. (39) can be rewritten as

$$f = \left| \int d^3x \varrho(\mathbf{x}) e^{-i\mathbf{k}_0 \cdot \mathbf{x}} \right|^2 \tag{41}$$

The recoil-free fraction is the square of the Fourier transform of the probability density. If $\varrho(\mathbf{x})$ is spread out over a large volume, f will be small. If, on the other hand, $\varrho(\mathbf{x})$ is concentrated in a small volume, f will be large. In the extreme case, we can assume that $\varrho(\mathbf{x})$ is a delta function; f then becomes unity. In the wave picture, these properties are easily interpreted. If the source nucleus moves over a large distance while radiating, the waves emitted from different points in space add up to a partially incoherent wave train. As a consequence the frequency of the wave is not well defined, and the energy is not sharp.

Equation (39) is very general. In order to get a more specific expression, we must use more detailed models. We begin with the simplest one and assume that the solid can be represented as a one-dimensional harmonic oscillator. The nucleus of mass M is then bound in this harmonic potential. Its total energy is the sum of the kinetic energy $p^2/2M$ and the potential energy $(M\omega^2/2)x^2$. Thus, the Hamiltonian of the system is

$$H = \frac{p^2}{2M} + \frac{M\omega^2}{2}x^2 \tag{42}$$

and any textbook on quantum mechanics shows how to find the energy eigenvalues

$$\varepsilon_n = \hbar\omega(\tfrac{1}{2} + n) \tag{43}$$

and the corresponding wave functions u_n. In particular, the ground-state wave function is given by

$$u_0(x) = \left(\frac{M\omega}{\pi\hbar}\right)^{1/4} e^{-(M\omega/2\hbar)x^2}, \qquad \varepsilon_0 = \tfrac{1}{2}\hbar\omega \tag{44}$$

Inserting this wave function into Eq. (39) yields with $k_0 = E_0/\hbar c$

$$f_0 = \left|\int_{-\infty}^{+\infty} dx\, u_0^*(x) e^{-i(E_0/\hbar c)x} u_0(x)\right|^2 = \exp\left\{-\frac{E_0^2}{2Mc^2\hbar\omega}\right\} = e^{-R/\hbar\omega} \tag{45}$$

With $\hbar\omega = E_E$, we have derived our earlier Eq. (29). The expression (45) can be rewritten into a more useful form. For the harmonic oscillator the average potential energy $\tfrac{1}{2}M\omega^2 <x^2>$ is equal to half the total energy $E_n = \hbar\omega(n + \tfrac{1}{2})$. Using his relation and $k_0 = E_0/\hbar c$, we can write for any level that

$$f = e^{-k_0^2 <x^2>} \tag{46}$$

This equation shows again that the Mössbauer effect is large if the emitting source is concentrated ("frozen").

Next we want to convince ourselves that on the average the oscillator still gains the recoil energy $R = \hbar^2 k^2/2M$. We calculate the total energy $<E_f>$ of the final state $\psi_f = e^{-ikx}u_i$

$$<E_f> = \int dx\, \psi_f^* H u_i = \int u_i^* e^{ikx}\left(\frac{p^2}{2M} + \frac{M\omega^2}{2}x^2\right)e^{-ikx}u_i dx$$

$$= \int\left[u_i^* e^{ikx}\left(-\frac{\hbar^2}{2M}\frac{\partial^2}{\partial x^2}\right)e^{-ikx}u_i + u_i^*\frac{M\omega^2}{2}x^2 u_i\right]dx$$

$$= \int u_i^*\left[\frac{\hbar^2}{2M}\left(k^2 - \frac{\partial^2}{\partial x^2}\right) + \frac{M\omega^2}{2}x^2\right]u_i dx = \frac{\hbar^2 k^2}{2M} + E_i$$

$$= R + E_{n_i}$$

We have used the fact that e^{ikx} commutes with any function of x, in particular, with the potential energy $(M\omega^2/2)x^2$, whereas it does not commute with the momentum operator $p = (\hbar/i)(\partial/\partial x)$. Thus we have verified that the average energy transferred to the oscillator is R as for a free particle.

In order to describe the behavior of an actual Mössbauer atom in a crystal, we have to generalize our treatment (1) to three dimensions and (2) to some 10^{22} modes of vibration in thermal equilibrium. The first generalization is trivial. We simply interpret $<x^2>$ of Eq. (46) as the mean-square displacement along the direction of the photon. We have to keep in mind, though, that $<x^2>$ may be large along one direction and small along another. Depending on the angle of emission with respect to this preferred direction we might observe different values for the recoilless fraction.

The second generalization is more involved. We still want to treat our source nuclei as harmonic oscillators, but we want to take a proper average over all possible modes of vibration that are occupied at a certain temperature T. We know from our previous discussion that $<x^2>$ is proportional to the energy: $<x^2> = \varepsilon/M\omega^2$.

If we remember from statistical mechanics that the average energy $<\varepsilon(\omega)>_T$ of a harmonic oscillator at temperature T is

$$<\varepsilon(\omega)>_T = \frac{\hbar\omega}{2} + \frac{\hbar\omega}{e^{\hbar\omega/kT} - 1} \tag{47}$$

we can immediately write down the thermal average of $<x^2>$

$$\ll x^2 \gg_T = \frac{<\varepsilon(\omega)>_T}{M\omega^2} \tag{48}$$

We can thus calculate the recoil-free fraction for an Einstein solid, which has only one vibrational frequency ω for all atoms, as discussed in the previous section.

We know, however, that the Einstein model is a crude approximation and in particular it fails to reproduce the low-temperature behavior of the specific heat. We therefore try the next better model, introduced by Debye, which allows all frequencies of vibration from $\omega = 0$ up to a maximum frequency ω_{max}. The latter is normally expressed in terms of the Debye temperature

$$\theta_D = \frac{\hbar\omega}{k} \tag{49}$$

Debye assumed that the number $N(\omega)$ of oscillators (phonons) of frequency ω is proportional to ω^2; if we normalize $N(\omega)$ to unity we have

$$N(\omega) = \frac{3\omega^2}{\omega^3_{max}} \tag{50}$$

Combining Eqs. (47), (48), and (50), we find for the average mean-square displacement

$$\langle \overline{x^2} \rangle = \frac{3\hbar}{\omega^3_{max} M} \int_0^{\omega_{max}} \left(\frac{1}{2} + \frac{1}{e^{\hbar\omega/kT} - 1} \right) \omega \, d\omega$$

or

$$\langle \overline{x^2} \rangle = \frac{3\hbar^2}{Mk\theta_D} \left[\frac{1}{4} + \left(\frac{T}{\theta_D} \right)^2 \int_0^{\theta_D/T} \frac{x \, dx}{e^x - 1} \right] \tag{51}$$

At very low temperatures the second term is small, and we can write

$$f_{(T=0)} = e^{-3E_0^2/4Mc^2k\theta_D} \tag{52}$$

whereas at temperatures $T \gtrsim \theta/2$ we can ignore the first term and expand the second to get

$$f_{(T \gtrsim \theta/2)} = e^{-[3E_0^2(T/\theta_D)]/(Mc^2k\theta_D)} \tag{53}$$

Typical Debye temperatures range from 80°K (CsI) to 440°K (Fe). The values scatter considerably depending on the type of measurement they are derived from, e.g., specific heat, elastic constants, neutron scattering, diffuse x-ray scattering, or recoilless fractions. This is not surprising since we know that the actual phonon spectrum $N(\omega)$, even for a simple solid, is much more complicated than the Debye spectrum $N(\omega)$=const ω^2. Nevertheless the last two equations provide a reasonable approximation for the recoilless fraction in many practical cases. Even if the Debye model is not strictly applicable, i.e., for noncubic lattices or for lattices with different masses, it provides a reasonable estimate.

Finally we mention another effect intimately related to the vibrational energy of a lattice, the thermal red shift or second-order Doppler effect. A photon emitted without recoil energy loss from a hot source has a lower energy than the same photon emitted from a cold source. The difference is readily measurable and has to be taken into account in precise measurements of energy shifts. In the hot source the mean-square velocity $\langle v^2 \rangle = E_n/M$ of a source nucleus is higher, and according to special relativity, the moving clock is seen to run slow by an outside observer. The source nucleus actually has its clock built in; it emits an electromagnetic wave of period $T=1/\nu$ $=h/E_r$. The period T_{hot} appears too long for an observer at rest, compared with his own period T_0. In fact, $T_{hot}/T_0 = 1/\sqrt{1-(v^2/c^2)}$, and the frequency ν_{hot} appears too low

$$\nu_{hot} = (1/T_{hot}) = \nu_0 \sqrt{1 - (v^2/c^2)} \cong \nu_0(1 - \langle v^2 \rangle/2c^2) \tag{54}$$

The thermal red shift is a demonstration of the famous twin paradox of special relativity.

We can arrive at the result [Eq. (54)] from a different argument. A mass M held by a spring with spring constant F has a frequency $\omega = \sqrt{F/M}$. If we decrease the mass by $E_0/c^2 \ll M$, the resulting frequency will be $\omega' = \sqrt{(F/M)[1/(1-E_0/Mc^2)]} \cong \omega[1+(E_0/2Mc^2)]$. If the quantum state of the oscillator has not changed, its energy will be $\varepsilon' = \varepsilon[1+(E_0/2Mc^2)]$. We conclude that of the total transition energy E_0, the amount $E_0(\varepsilon/2\ Mc^2) = E_0(M<v^2>/2Mc^2)$ is lost to the vibrational energy; therefore, the photon carries off an energy $E_0[1-(<v^2>/2c^2)]$. Clearly $<v^2>$ is a function of temperature if the harmonic oscillator is in thermal equilibrium with its surroundings.

REFERENCES

1. V. Weisskopf and E. Wigner, *Z. Physik* **63**, 54 (1930); **65**, 18 (1930). See also, W. Heitler, *Quantum Theory of Radiation* (Oxford University Press, London, 1954).
2. Such experiments have been performed very successfully. A survey is given by K. G. Malmfors, in *Alpha-, Beta-, and Gamma-Ray Spectroscopy*, K. Siegbahn, Ed. (North-Holland, Amsterdam, 1965), Vol. 2.
3. A. Einstein, *Ann. Physik* **22**, 180 (1907).
4. P. Debye, *Ann. Physik* **39**, 789 (1912).
5. M. Born and T. von Karman, *Physik. Z.* **13**, 297 (1912).
6. R. L. Mössbauer, *Z. Physik* **151**, 124 (1958); *Naturwiss.* **45**, 538 (1958); *Z. Naturforsch.* **14a**, 211 (1959).
7. H. J. Lipkin, Physics 480, University of Illinois, unpublished.

Chapter 2

Instrumentation

Jon J. Spijkerman

National Bureau of Standards
Washington, D.C.

During the first decade since the discovery of the Mössbauer effect, instrumentation and techniques have been developed to a high degree of sophistication. The present instrumentation is the result of many innovations by the researchers in this field, and commercial spectrometers now available are based upon their design. Much work has been done in the development of associated equipment to study Mössbauer sources or absorbers at variable temperatures in an applied magnetic field or at high pressures. Procedures for making sources are well documented for many isotopes, and for the more popular isotopes the sources are available commercially. Backscattering techniques have reduced the problem of sample preparation and opened the way for possible commercial applications. However, the time required to obtain a spectrum is still relatively long, even with high-speed counting systems, and the data processing requires a computer, particularly for the more complicated spectra.

1. INSTRUMENTATION

The earlier type of Mössbauer spectrometers used mechanical means to obtain the Doppler motion, generally in the constant-velocity mode. Although these mechanical units are simple and have good reproducibility, they are difficult to interface for an automated system. The trend has been towards electromechanical spectrometers coupled with a multichannel analyzer in the constant acceleration mode. Most commercial spectrometers are of this latter type, and the principles of their operation will be discussed. The spectrometer consists of two main parts, the gamma-ray detection and recording system and the Doppler velocity drive.

1.1. Gamma-Ray Detection

Since most Mössbauer sources are not monochromatic and emit radiation of higher or lower energy than the Mössbauer gamma ray, the detection system must count only the Mössbauer radiation. A typical energy spectrum of a ^{57}Co source is shown in Figure 1. Three types of detectors are used in Mössbauer spectroscopy—the proportional counter, the scintillation detector, and the solid-state detector.

The proportional counter is generally used for the energy region from 1 to 20 keV, and this type of counter is used extensively for ^{57}Fe Mössbauer spectroscopy. The counter consists of an outer cylinder at ground potential and a center wire anode at a high positive potential. The counter is filled with either argon, krypton, or xenon. A gamma ray entering the counter ionizes the gas and forms ion pairs. The electrons will accelerate to the anode and form other ion pairs by collision with the gas atoms. Multiplication factors as high as 10^5 can be obtained, and the anode current will be proportional to the gamma-ray energy. To prevent a continuous electrical discharge in the counter, a quenching gas, such as methane, is added, which dissipates the energy by dissociation. Commercial counters once had a lifetime of 10^9

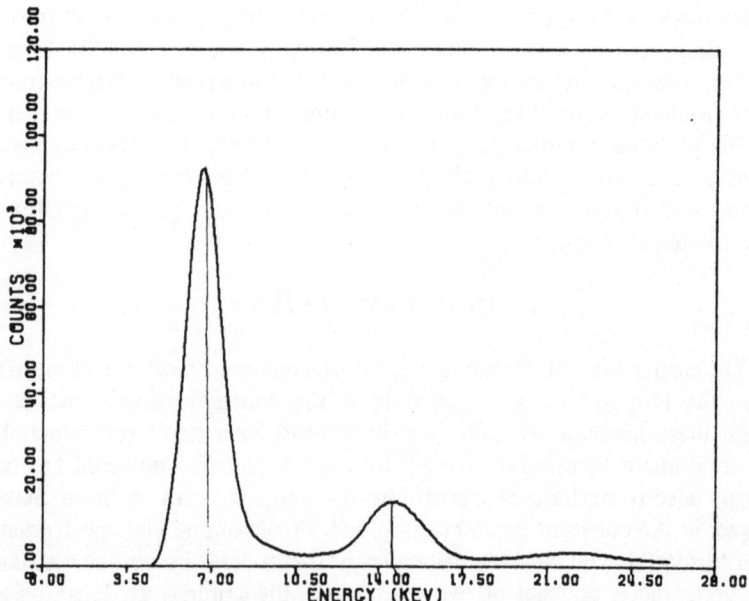

Figure 1.   Energy spectrum of ^{57}Co with a proportional counter (Kr-CO$_2$ gas mixture).

counts before the quenching gas became depleted. However, high-gain solid-state amplifiers reduce the requirement for large gas multiplication factors, and with recently innovated CO_2 quenching gas the counter life is now lengthened. For ^{57}Fe spectroscopy, a 2-in.-diameter, 1-atm krypton-filled counter is recommended. This type of counter is about 60% efficient for gamma radiation and has 12% resolution.

For higher gamma-ray energies, such as the 23.8-keV ^{119}Sn Mössbauer radiation, the scintillation counter is used. This detector consists of a thin thallium-doped NaI crystal mounted on a photomultiplier. The gamma-ray energy is converted by the crystal to visible light, and these photons produce a current in the photomultiplier. A 2-mm NaI crystal is about 97% efficient for the 23.8-keV ^{119m}Sn radiation, but the energy resolution is at best 20%.

The solid-state detector has extremely good energy resolution (600 eV at 14.4 keV), but it must be cooled below 120°K. It consists of a PIN (p type, intrinsic, n type) semiconductor made of lithium-doped silicon or germanium. The intrinsic region is normally nonconducting, but when the gamma radiation ionizes the lithium-doped region, conduction takes place and a pulse is produced. The pulse height is proportional to the gamma-ray energy. The cost of these detectors has prevented general use in Mössbauer Spectroscopy.

The schematic for this type of gamma-ray detection system is shown in Figure 2. Since the output impedance of the detector is very high, a preamplifier is used, followed by a linear amplifier. The peak voltages of the pulses produced by the linear amplifier are proportional to the gamma-ray energy. The single-channel analyzer (SCA) now selects only pulses above a threshold, but in the "window" of the SCA, so that the gamma-ray energy can be selected. The simplest method of adjusting the SCA is by using the coincidence circuit of the multichannel analyzer in the pulse-height mode.

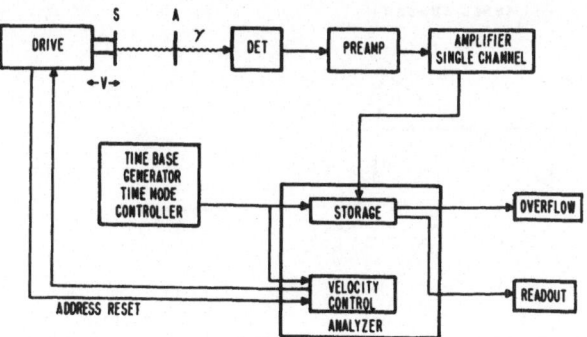

Figure 2. Block diagram of Mössbauer spectrometer.

1.2.   Doppler Velocity Drive

A multichannel analyzer coupled with a Doppler spectrometer presents two methods of conveniently obtaining Mössbauer effect data in the constant-acceleration mode. The first is the modulation of the pulse height by the driving waveform [1] with the analyzer in the pulse-height analysis mode. This method will give a good Mössbauer spectrum if the analog-to-digital converter (ADC) is linear, but the count rate is restricted because of the long dead time required to analyze and address each pulse into the memory. The second method is to run the analyzer in a multiscaler mode and derive a constant increment of velocity for each channel. This method is superior to the first because the dead time is no longer a factor related to channel number of the memory and the linearity requirement of the ADC is eliminated. However, the accuracy of the spectrum produced by this method is only as good as the synchronization between the analyzer and the spectrometer. The analyzer can generate either a triangular or sawtooth waveform. The triangular waveform is generated by integration of the square wave [2] obtained from the switching of the analyzer subgroups. The motion of the drive is a double parabola, and two spectra, mirror images, are displayed on the cathode-ray screen. The sawtooth waveform is directly obtained from the analog voltage of the address scaler [3]. The motion is a single parabola with flyback (Figure 3) and a single spectrum is displayed. The decoding of

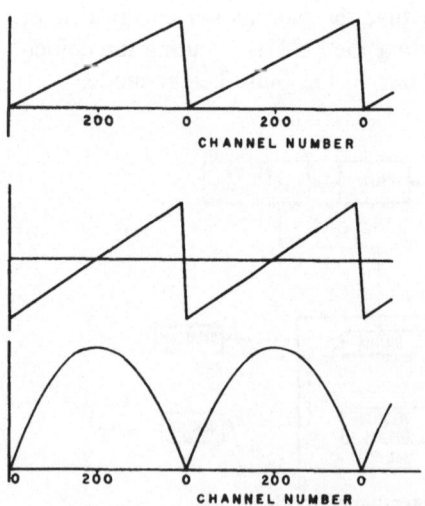

Figure 3.   Waveforms of input voltage (top), velocity (middle), and displacement of the drive (bottom).

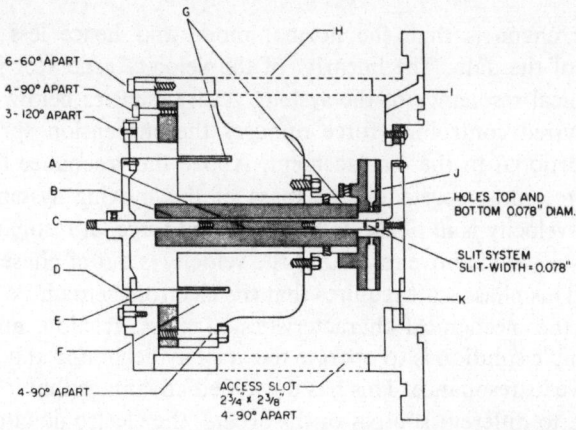

Figure 4. Cutaway view of velocity transducer. Permanent magnet A of linear velocity transducer B is rigidly connected to drive shaft C, Flexure plates D and K support loudspeaker coil E with permanent magnet F. Photocell J provides limits to the motion by means of the slit system. From *NBS Tech. Note* No. 421 (1967) p. 7.

the address scaler provides continuous synchronization of the analyzer and the drive unit.

The most widely used and commercially available velocity generators are based upon an electromechanical feedback system, shown in Figure 2. The driving force is produced by a current through a coil in a permanent magnet, mechanically coupled to a linear velocity transducer, which provides the feedback signal (see Figure 4). Since the feedback is magnetically coupled, the direct current level must be maintained, or drift will occur. Stiff suspension springs can be used to minimize this effect, but this will raise the natural resonance of the velocity generator. By using a photocell and slit to trigger each sweep, the drift can be eliminated, as it is with a chopper-stabilized amplifier. The photocell–slit system can also be used in a constant-velocity mode, where the required input signal is a rectangular waveform. The velocity generator produces this function from a photocell and double slits, which triggers a bistable circuit [4]. This circuit can be designed so that the symmetry of the rectangular wave can be changed, to obtain different velocities in the forward and backward directions.

The feedback system makes it possible to generate any desired velocity waveform from the corresponding voltage signal. Besides the square, triangular, and sawtooth waveforms, a trapezoidal function is used for a "region-of-interest" type of spectrometer. The drive quickly accelerates to a high velocity and then slowly sweeps over the desired velocity range. This requires fewer channels than the normal mode and hence less time for accumulation of the data. The linearity of the velocity generator is limited by the mechanical resonance of the system. At frequencies below the resonance the required controlling force opposes the suspension springs and hence is proportional to the displacement. Above the resonance frequency the driving force must accelerate the mass of the moving system. Below resonance, the velocity is in phase with the applied force, at resonance there is a 90° phase shift, and above resonance the velocity is out of phase with the applied force. This phase shift requires that the electronic circuit be carefully matched with the mechanical characteristics of mass, friction, and spring constant. A simple solution is to operate the velocity generator at a scanning frequency above its resonance. This has the added advantage that if the mass is changed due to different sources or absorbers, the electronic circuit does not require adjustment. However, a high scanning frequency has several drawbacks. To reproduce the velocity waveform with high fidelity from the input voltage, a Fourier analysis shows that higher harmonics must be present. Since the frequency limit of these mechanical generators is about 10 kHz, scanning frequencies above 10 Hz are not recommended. To obtain a low scanning frequency requires large displacements to obtain the desired velocity, which introduces distortion in the baseline of the spectrum. Recommended scanning rates are between 1 and 6 Hz, depending on the velocity range.

2. ACCURACY AND PRECISION

The degree of precision to which a Mössbauer parameter can be measured is limited by the vibration, nonlinearity, and drift in the Doppler drive, which can be determined experimentally. Instrumental errors, due to the Doppler motion, are:

1. *Parabolic distortion due to the inverse-square law*, which is particularly noticeable for a moving source and small source–detector separation. The parabolic distortion of the baseline can be eliminated by using a moving absorber.
2. *Compton scattering*, when higher energy radiation is emitted by the source. Even for a moving absorber, this will give a parabolic baseline distortion.
3. *Cosine smearing*, which produces a spectrum shift toward higher velocities and is a direct consequence of the angular dependence of the Doppler effect. The velocity shift is $dv = vD^2/16d^2$, where D is the detector diameter and d the source detector distance [5,6].
4. *Channel-width broadening*, due to the velocity change while counts

are accumulated in a channel. This depends on the calibration constant (mm/sec/channel).

2.1. Calibration Methods

A spectrometer can be calibrated by a direct velocity measurement of distance traversed in a unit of time for constant-velocity spectrometers, or by using a known Mössbauer spectrum for comparison, as is usually the case for constant-acceleration spectrometers.

The National Bureau of Standards has issued standard reference materials for this purpose [7].[1] Optical interferometric techniques provide an absolute method. By counting the fringes from a Michelson interferometer with a laser light source [8,9], or using a moiré grating [10] with a multichannel analyzer, the velocity for each channel can be obtained. The interferometer can also be included in the feedback loop for very accurate measurements and long-time stability.

The Michelson interferometer makes use of the interference between two coherent light beams. A diagram of this interferometer is shown in Figure 5a. The coherent light from a laser is split by means of a beam splitter into two equal-intensity beams, which are reflected back by a stationary and a moving mirror, respectively. Interference takes place when the two beams are recombined by the beam splitter, and the light intensity will change from bright to dark when the path difference is one-half wavelength. Since the wavelength of an argon–neon laser is 6328.198 Å and the path is traversed twice, a displacement of 1582.05 Å will generate an electrical pulse by the photodetector, which can be counted by the analyzer. In this mode, the multichannel analyzer will display the true velocity waveform. The velocity

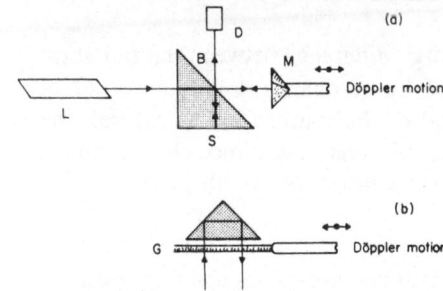

Figure 5. (a) Diagram of Michelson interferometer with laser L, beam splitter B, stationary mirror S, moving corner cube M, and photocell D. (b) Moiré fringe system. Light from lamp I pases through grating G through prism back to photocell D.

---

[1] Sodium nitroprusside has been issued as a reference standard for isomer shift of iron compounds; an iron-foil velocity standard and a tin standard (dimetyl tin difluoride) will be issued by the NBS in September 1970.

is calculated from the number of fringes in each channel, the sweep frequency, and the number of sweeps, or by counting the frequency of an oscillator in channel one simultaneously with the interference fringe counting in the other channels.

The moiré grating technique uses a ruled grating as a light shutter. The light is passed from a source through the grating and reflected back to a photodetector. The moiré system is much simpler to align than the interferometer, but is an order of magnitude less sensitive than the interferometer; this is, however, sufficient for Mössbauer spectroscopy.

2.2. Time of Counting

The increased accuracy of the available spectrometers makes it necessary to estimate the time required to obtain the Mössbauer spectrum parameters to the maximum accuracy of which the instrument is capable. Protop and Nistor [11] have calculated the standard error in the position expected for a single resonance line

$$\sigma(v) = \frac{2\sqrt{2\Gamma_{exp}}}{\varepsilon\sqrt{\pi n}\sqrt{N}} \quad \text{with } n\Gamma_{exp} > 4 \quad (1)$$

where Γ_{exp} is the experimental half width of resonance line in mm/sec, ε is the resonance effect magnitude, n is the calibration constant (channels/mm/sec), and N is the number of counts in the baseline. The effect can usually be obtained from the analyzer cathode-ray screen after a short period of data accumulation. The final error estimates are generally obtained from the computer program.

3. EXPERIMENTAL TECHNIQUES

In Mössbauer spectroscopy the energy difference between the radiation emitted by the source and that required for resonance is only an order of magnitude larger than the energy spread of the source (due to natural line width). This low resolution and the generally small resonance effect requires great care in source preparation and optimization of the absorber.

3.1. Sources

Mössbauer sources can be divided into two groups, depending upon the radioactive decay of the parent nucleus. For an isomeric transition, such as found in ^{119m}Sn and ^{125}Te, the source and absorber can be chemically alike. This is not the case for a β emitter or electron capture transition where a suitable matrix material must be found. Selection of the matrix is governed by (a) its Debye temperature, (b) atomic attenuation of the radiation, (c) production of x-ray fluorescence, and (d) lattice parameters in order to

Table 1. Mössbauer Sources[a]

Absorbing nucleus	Parent	Half-life	γ-ray energy, keV	X-ray energy, keV	Filter, mils	Source
^{57}Fe	^{57}Co	270 days	14.4125	6.5	5 Al	^{57}Co in Pd
^{61}Ni	^{61}Co	1.7 hr	67.4	7.6	—	^{62}Ni (18%Cr)
^{83}Kr	^{83m}Kr	2.4 hr	9.3	12.8	—	Kr–Clathrate
^{119}Sn	^{119m}Sn	250 days	23.8	25.8	2 Pd	Ba^{119m}SnO$_3$
^{121}Sb	^{121m}Sn	5 yr	37.2	26.9	—	Sn metal
^{125}Te	^{125m}Te	58 days	35.5	28.0	—	^{125}Sb in Cu
^{127}I	^{127}Te	9 hr	57.6	29.2	—	ZnTe
^{129}I	^{129}I	70 min	27.75	29.2	4 In	ZnTe
^{129}Xe	^{129}I	10^7 yr	39.6	30.4	—	NaI
^{149}Sm	^{149}Eu	106 days	22	41.0	—	Eu$_2$O$_3$
^{151}Eu	^{151}Sm	120 days	21.6	42.5	—	SmF$_3$, Sm$_2$O$_3$
^{161}Dy	^{161}Tb	6.9 days	25.7	47.0	—	Gd$_2$O$_3$
^{169}Tm	^{169}Er	9.3 days	8.41	9.48, etc.	—	Er$_2$O$_3$
^{195}Pt	^{195}Au	192 days	98.8	68.4	—	^{195}Au in Pt
^{197}Au	^{197}Pt	20 hr	77.3	70.4	—	Pt metal
^{237}Np	^{241}Am	458 yr	59.5	103.5	—	5% Am in Th

[a] A. H. Muir, Jr., K. J. Ando, and H. M. Coogan, *Mössbauer Effect Data Index*, 1958–1965 (Interscience, New York, 1966); J. J. Spijkerman in *Technique of Inorganic Chemistry*, H. B. Jonassen and A. Weissberger, Eds. (Interscience, New York, 1968), Vol. 7; and J. R. DeVoe and J. J. Spijkerman, *Anal. Chem.* **40**, 472R (1968).

minimize quadrupole and magnetic interactions. The source matrixes for several Mössbauer nuclei are listed in Table 1. No proven formula exists for the determination of a suitable matrix, but the host lattice should have a high Debye characteristic temperature θ_D, since this will determine the recoil-free fraction, f. This fraction can be calculated from the Debye–Waller equation [12], or

$$f = \exp[3E_0^2/4Mc^2k\Theta_D] \qquad (2)$$

where E_0 is the photon energy, M is the atomic mass, and k is the Boltzmann constant. For an impurity atom in a host lattice, the effective Debye temperature is

$$\Theta'_D = (M_{host}/M_{impurity})^{1/2}\Theta_D \qquad (3)$$

where M_{host} and $M_{impurity}$ are the atomic mass of the host lattice atoms and the impurity (source) atoms. The Debye temperatures for the elements and some compounds have been reported by Holm [13]. For most compounds the Debye temperatures are not available, but a good estimate can be obtained from the specific heat data since the Debye temperature is also defined as the temperature for which the specific heat of a crystal takes on the value of 5.670 cal/mole-deg.

The use of filters to attenuate non-Mössbauer radiation can greatly increase the efficiency of the detector. In the case of ^{57}Fe, where the 6.3-keV x ray is an order of magnitude more intense that the 14.4-keV Mössbauer radiation, a 0.005-in. aluminum foil will decrease the x-ray intensity by a factor of 50, while the 14.4-keV radiation attenuation is only 3%. For ^{119m}Sn the palladium absorption edge is used to filter the 23.8-keV radiation from the 25.8-keV x rays. Suitable filters are listed in Table 1.

3. 2. Mössbauer Absorbers

The thickness optimization and sample mounting are the main considerations in absorber preparation. The thickness optimization has been studied in detail by Shimony [14] and Protop and Nistor [11]. The optimum absorber thickness is a function of (a) the atomic absorption, (b) the number of Mössbauer nuclei per unit area, and (c) the production of scattered radiation in the absorber. The attenuation of the gamma radiation by the absorber can be expressed by

$$I/I_0 = e^{-\mu x} = e^{-(\mu/\rho)(\rho x)} \tag{4}$$

where μ/ρ is the mass absorption coefficient in cm^2/g and ρx is the sample thickness in g/cm^2. The attenuation coefficients for the 14.4- and 23.8-keV radiation are shown in Figure 6a and b. The Mössbauer resonance effect magnitude can be calculated from

$$\varepsilon = \frac{I_\infty - I_0}{I_\infty} = f_s(1 - e^{-T/2}J_0(iT/2)) \tag{5}$$

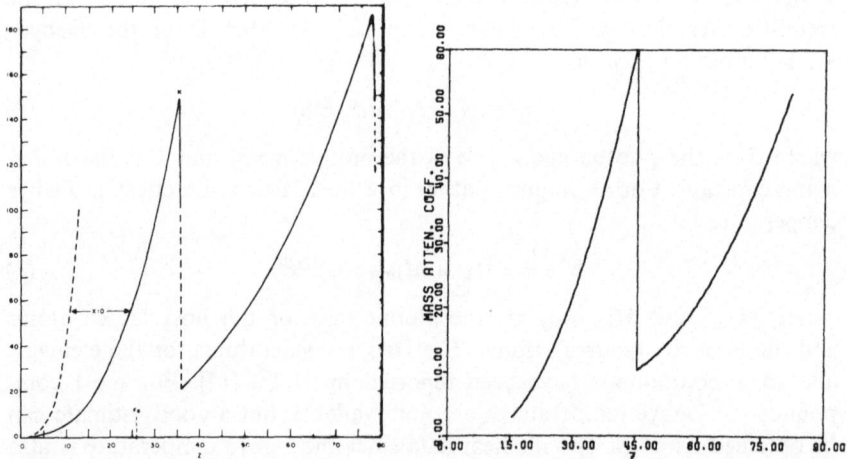

Figure 6.   Left, attentuation coefficient for 14.4-keV gamma radiation as a function of the atomic number of the absorber material. Right, attenuation coefficient for 23.8-keV gamma radiation as a function of the atomic number of the absorber material.

where

$$T_a = \sigma_0 f_a an$$

and f_s, f_a are the source and absorber f factors, σ_0 is the resonant cross section, a is the abundance of the Mössbauer isotope, and an is the number of Mössbauer nuclides per unit area. Too thick an absorber will cause line broadening, with the broadening given by

$$\Gamma_{exp} = (2 + 0.27T)\Gamma_{nat} \tag{6}$$

where Γ_{nat} is the natural line width.

The Compton scattering is difficult to calculate for different absorbers. In general the limiting factor is the attenuation, and 30% transmission can be used for an initial trial as judged from the pulse-height spectrum. This spectrum also gives a good measure of the Compton scattering, and the Mössbauer peak should be at least 20% above the background in the pulse-height spectrum. Equation (6) can then be used to estimate the thickness broadening and the need for reducing the absorber thickness. The geometrical position of the absorber is not too critical. Since the internal conversion coefficients for iron and tin are large, the probability of detection of the reemitted Mössbauer radiation by the absorber is small. If the Compton scattering produced by the absorber is large, the absorber should be placed closer to the source to reduce the inverse-square-law distortions.

The technique and materials for mounting the absorber have been described by May and Snediker [15]. Plexiglass and polyethylene are the most suitable materials, provided their iron content is low, for mounting powder samples. Teflon and nylon are also used, particularly at low temperatures. The half-thickness (50% attenuation of the 14.4-keV gamma radiation) is 7 mm for plexiglass and 8.4 mm for polyethylene. Plasticizers and epoxy resins are also available for mounting powder samples permanently. For high-temperature work the samples can be clamped between beryllium disks, but the iron content of beryllium is generally high.

3.3. f-Factor Measurements

The f-factor measurements are difficult to perform. The most direct method is by measuring the effect for an identical source and absorber and using Eq. (4) or a series of absorbers of known concentration, as is shown in Figure 7. Relative measurements of f factors for sources can be made by using a "black absorber" in the form of fluoroferrates, as described by Housley et al. [16]. This absorber has a composite line width of 1.4 mm/sec. The accuracy in f determinations are limited by [16] (a) resonant self-absorption in the source, (b) nonresonant absorption in the source and absorber, (c) scattering of gamma radiation by the source and the absorber,

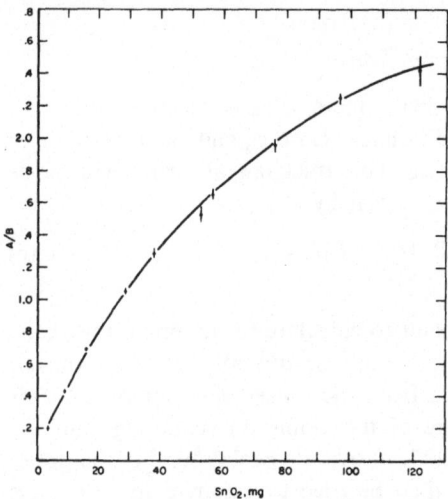

Figure 7. Mössbauer absorption as a function of SnO_2 concentration. Peak absorption area A to baseline B, Ratio *vs* mg SnO_2. From Pella *al.*, *Anal. Chem.* **41**, 46 (1969).

and (d) reemission of resonant radiation by the absorber. For [57]Co in a platinum matrix, $f(Pt) = 0.736 \pm 0.023$ [17].

3.4. Variable Temperature

Since most Mössbauer isotopes require low temperatures to observe the effect, and temperature dependence of the Mössbauer parameters is often required for characterization of a material, the present trend in auxiliary instrumentation is to cover a temperature range of 1.2 to 1000°K. Dewars for a temperature range of 1.2 to 300°K have been described in detail by Kalvius [18] and Benczer-Koller and Herber [19]. Some recent designs will be discussed because of their simplicity and economy. Liquid nitrogen is adequately suited for a coolant in the 80–300°K range. A simple cryostat can be constructed by using a cold-trap resevoir and sample holder as shown in Figure 8. The styrofoam does not attenuate the gamma radiation and is an excellent insulator. A cooled source and absorber modification has been described by Travis and Spijkerman [20], with a liquid nitrogen consumption of 0.2 liters/hr.

Liquid helium is used as a coolant below 80°K. Stainless steel research dewars with beryllium windows are commonly used, but the helium consumption is rather high. Glass dewars are much more economical, but until recently the windows were a problem. Epoxy-coated Mylar[2] is now available which will bond to glass and withstand the low temperatures encountered.

---

[2] G. T. Schjeldahl Corporation, Northfield, Minnesota.

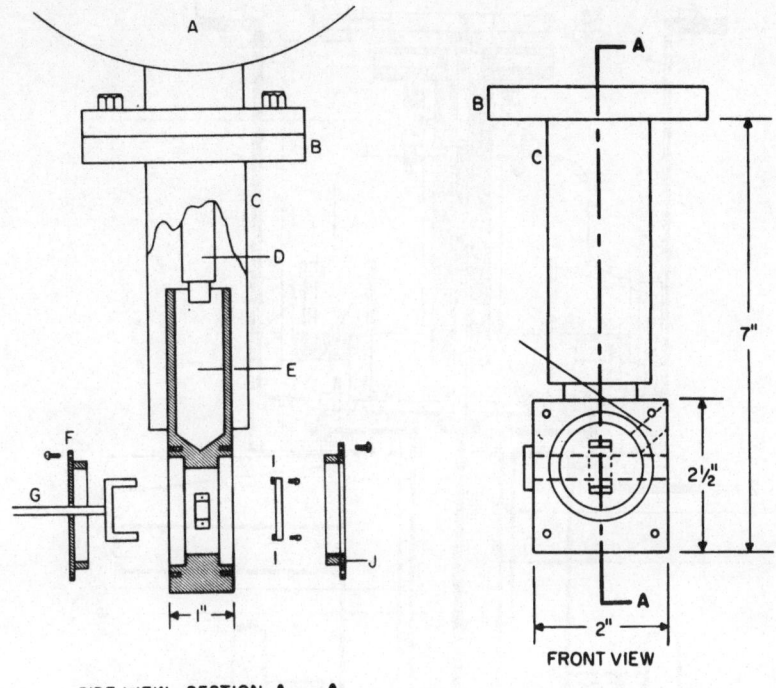

SIDE VIEW—SECTION **A——A**

Figure 8.   Simple crystat for moving absorber and stationary source.

The temperature can be varied by using an exchange gas in the sample chamber and a heating coil, as shown in Figure 9.

Temperatures can be conveniently measured with a gold (doped with 0.07 at. % Fe)–chromel thermocouple and a potentiometer over the range of 4.2 to 300°K. The thermoelectric voltage of this thermocouple as a function of temperature is shown in Figure 10. For accurate temperature measurements, carbon, germanium, or platinum resistance thermometers should be used [21]. The temperature range for the carbon resistor is 0.1 to 20°K, for the germanium resistor 4.2 to 100°K, and for the platinum resistor 20 to 300°K.

The use of high-temperature ovens for Mössbauer work [22,23] has received much less attention than cryogenic equipment. Electrically heated resistance ovens can be modified, but the geometry is in general not very good. A simple oven can be constructed by using a carbon-cloth heater [24]. The carbon cloth[3] can be heated to very high temperatures, and the sample

---

[3] Union Carbon and Carbide Corporation, Oak Ridge, Tennesee.

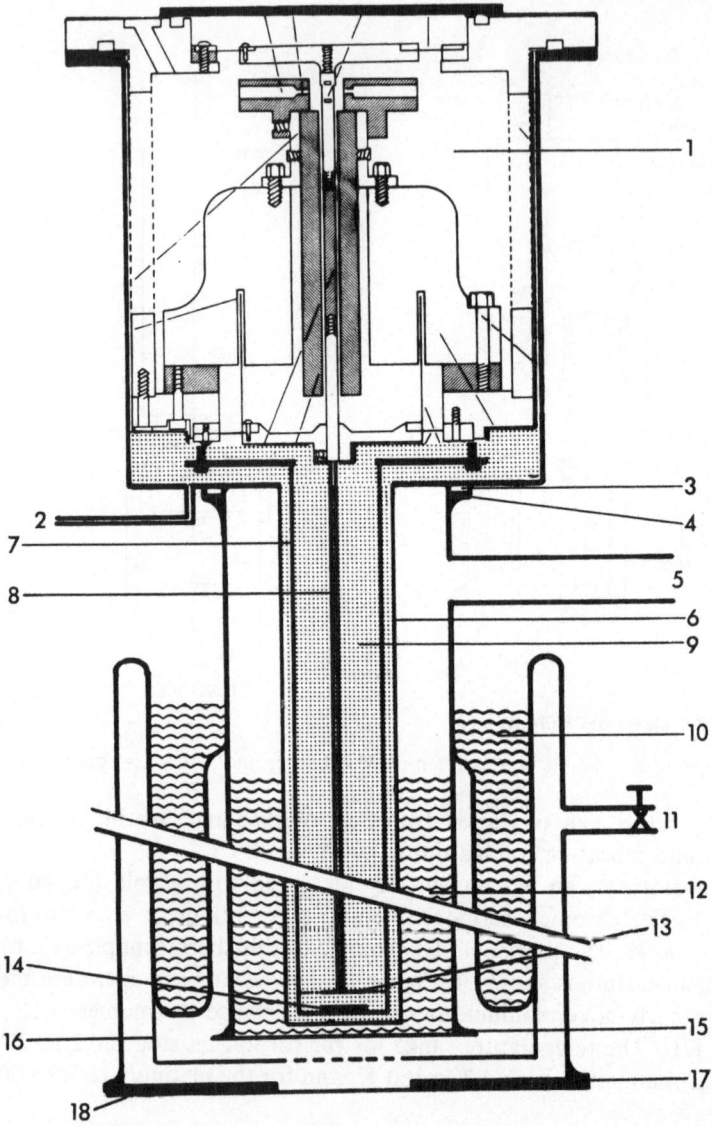

Figure 9.   Helium-temperature glass cryostat. (1) Velocity drive unit, (2) helium exchange gas inlet, (3) teflon gasket, (4) ground glass surface, (5) pumping arm, (6) stainless steel heat exchanger tube, (7) stainless steel tubing for absorber support, (8) stainless steel pushrod, (9) helium exchange gas, (10) liquid nitrogen, (11) high-vacuum valve for seal-off, (12) liquid helium, (13) source supported by flexure plate. (14) absorber and heating coil, (15) ground glass flange, with epoxy-coated Mylar seal, (16) metal heat shield, (17) ground glass flange for O-ring seal, (18) metal bottom flange with Mylar exit window. The velocity drive unit and source-absorber mounting can be lifted out of the cryostat for changing absorbers.

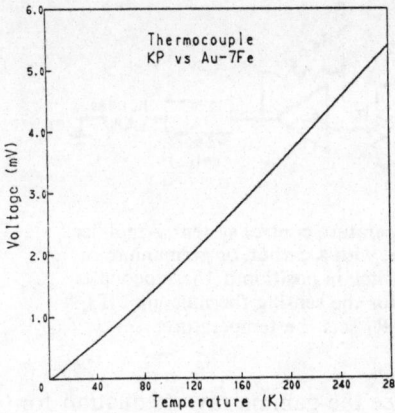

Figure 10. Thermoelectric voltage of chromel *vs* gold–0.07 at.% iron. From *NBS Rept.* No. 9712 (1968).

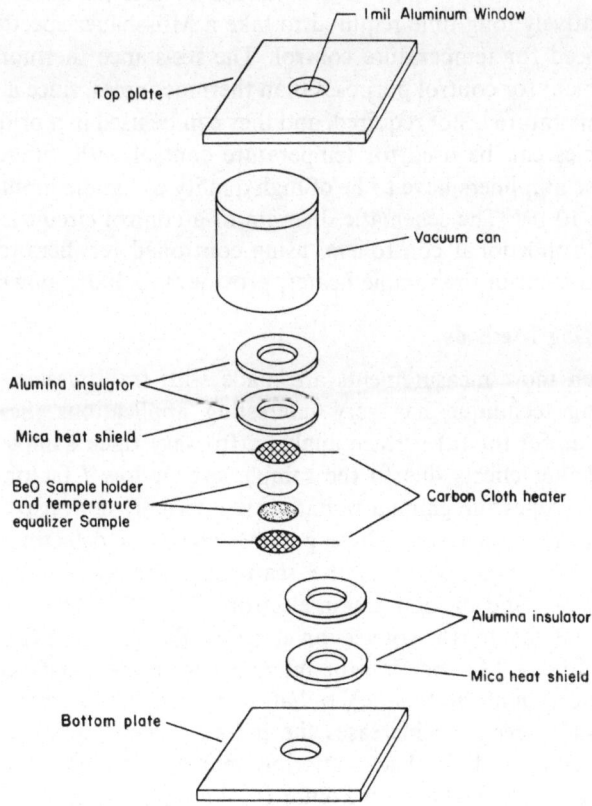

Figure 11. Diagram of high-temperature oven using carbon-cloth heating element.

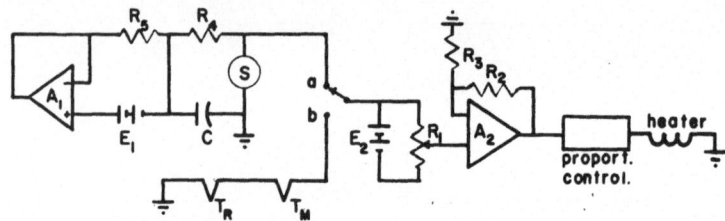

Figure 12. Diagram of proportional temperature control system. Amplifier A₁ is used for a constant current generator, with a carbon or germanium resistor for temperature sensor S. With the switch in position b, thermocouples can be used. T_R is the reference junction for the sensing thermocouple T_M. Amplifier A₂ must be chopper stabilized. R₁ sets the temperature.

placed between two heating elements, since the gamma-ray attenuation for the carbon cloth is very low. A Pt–Pt(0.1% Rh) thermocouple can be used to measure the temperature. A diagram of this oven is shown in Figure 11.

The relatively long time required to take a Mössbauer spectrum necessitates the need for temperature control. The resistance thermometers are more convenient for control purposes than thermocouples, since a reference-junction temperature is not required, and they can be used in a bridge circuit. Thermocouples can be used for temperature control with differential amplifiers. These amplifiers have to be of high quality to handle input signals in the order of 10 μV. The schematic diagram of a control circuit is shown in Figure 12. Proportional controllers, using controlled rectifiers to gate the alternating current in the sample heater, provide 0 to 100% power control.

3.5. Scattering Methods

Although most measurements are made with transmission geometry, the scattering technique has very interesting applications. Scattering is particularly useful for (a) surface analysis, (b) very thick samples, and (c) small Mössbauer effects due to the sample size or low f factor. In many Mössbauer isotopes the gamma radiation is internally converted into an x ray and conversion electron. Three possible means of detecting the Mössbauer effect are then possible: the scattered Mössbauer radiation, the conversion x ray, and the conversion electron. For ⁵⁷Fe, the x-ray energy is 6.3 keV and 7.3 keV for the conversion electron. The geometry for scattering is shown in Figure 13. For ⁵⁷Fe the internal conversion coefficient is nine, hence the detection of the 14.4-keV radiation is inefficient. x-ray fluorescence by the 122-keV precursor increases the noise for x-ray detection. A proportional counter with 2π backscattering geometry designed by Swanson and Spijkerman [25] is shown in Figure 14. An argon–10% methane gas is the most efficient for x-ray detection, while the conversion electrons can be detected by a helium–10% methane mixture without x-ray interference.

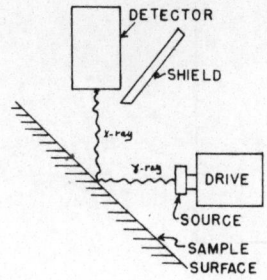

Figure 13. Geometry for backscattering.

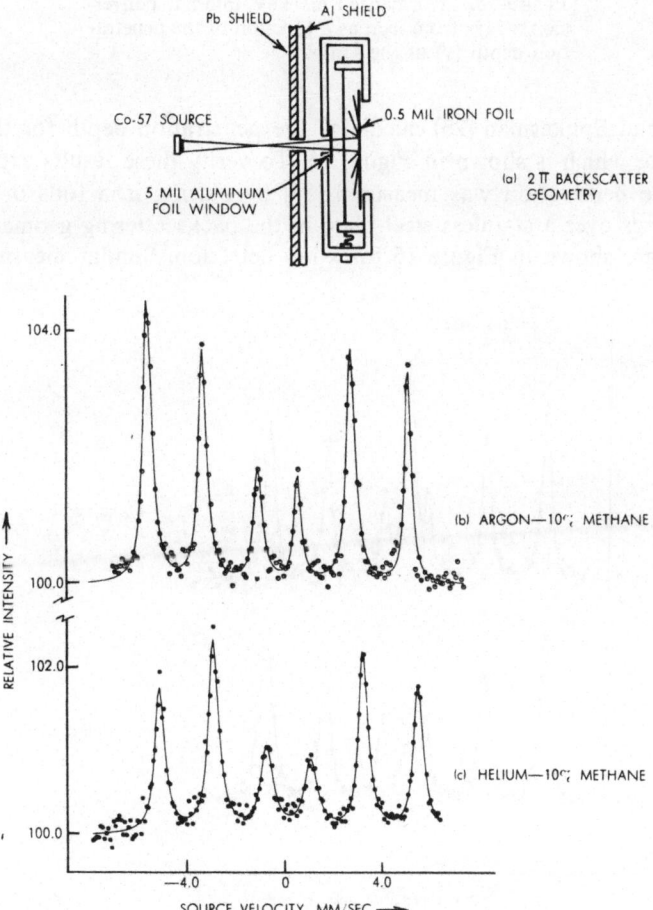

Figure 14. (a) Proportional counter with 2π geometry, (b) spectrum of iron foil using argon-10% methane counting gas, (c) spectrum of iron foil using helium-10% methane counting gas.

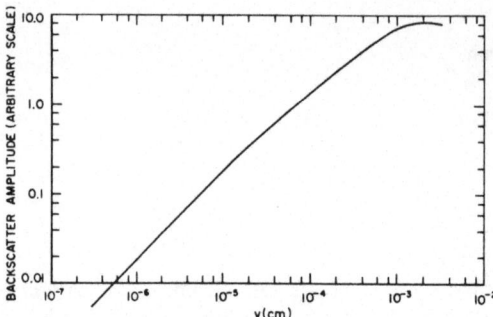

Figure 15. Normalized 6.3–keV internal conversion x rays from iron as a function of the penetration depth (y) of the sample.

Terrell and Spijkerman [26] calculated the penetration depth for the x-ray detection, which is shown in Figure 15. To verify these results experimentally, the penetration was measured [25] by placing iron foils of various thicknesses over a stainless steel plate in the backscattering geometry. The spectra are shown in Figure 16 for x-ray detection. Similar measurements

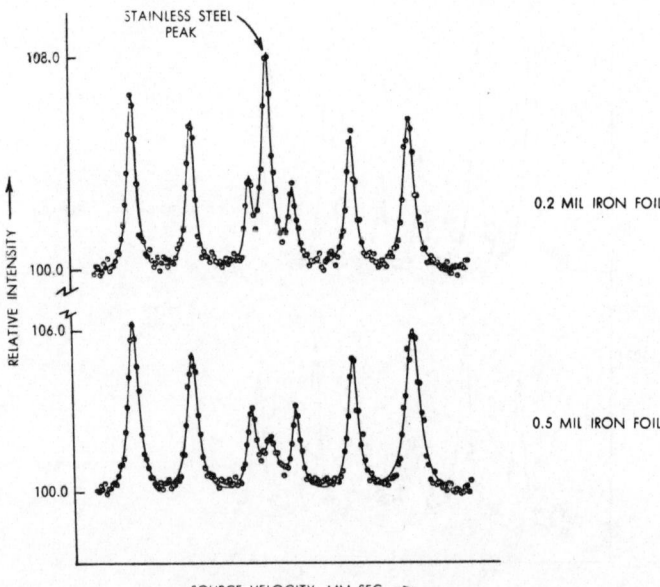

Figure 16. X-ray Mössbauer backscattering spectra for iron foils of two thicknesses on stainless steel foil.

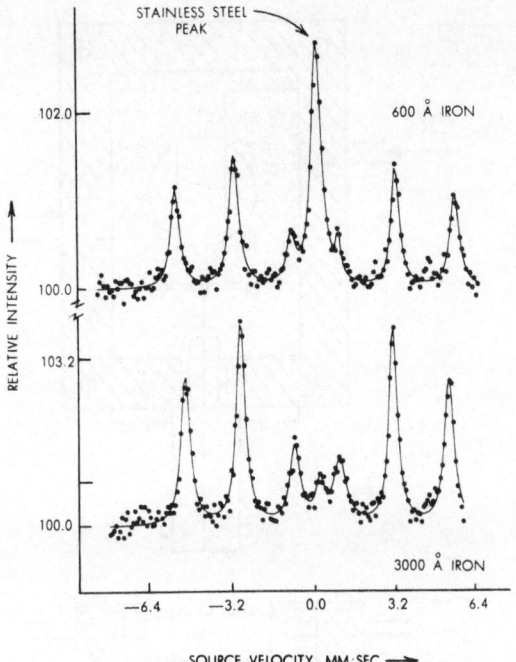

Figure 17. Conversion electron Mössbauer back-scattering spectra for vacuum-deposited iron of two thicknesses on stainless steel.

were made using electron detection, as shown in Figure 17. The penetration of the 7.3-keV electrons is limited to 3000 Å, which makes this an ideal technique for surface analysis. If an enriched stainless steel foil is used, this counter can be used as a resonance detector. Mössbauer effects of 400% with a 100% baseline have been obtained [27]. The design of such a detector is shown in Figure 18.

4. APPLICATIONS OF COMPUTERS TO MÖSSBAUER SPECTRA

Several computer programs are now available for Mössbauer data processing. These programs can be classified in three main categories: (a) computation of Mössbauer spectra from a theoretical model; (b) curve fitting of Mössbauer data by least-square analysis; and (c) curve fitting of experimental data by a constraint least-square analysis. Most of these programs are written for medium-size computers, and with the time-sharing computers (remote access by telephone line) now available the problem of data analysis is greatly reduced. It is interesting to note that all programs

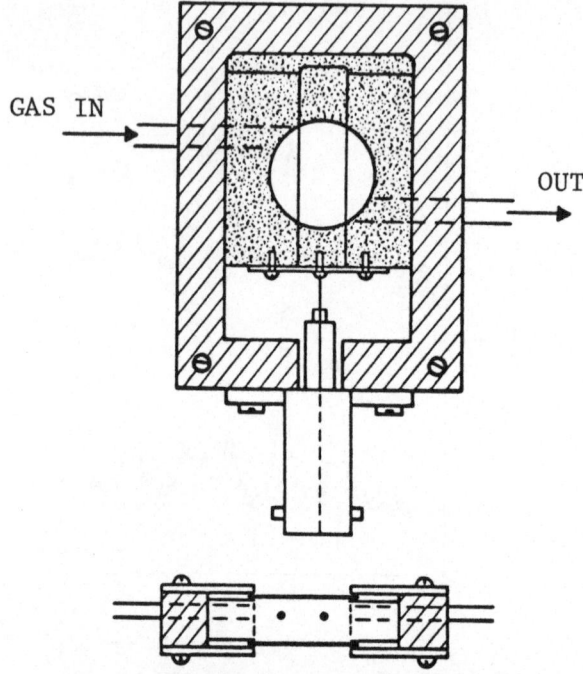

GAS IN

OUT

Figure 18. Design of ^{57}Fe resonance detector.

written use a Lorentzian line profile. Experimental data quite often shows a deviation of this profile, and a better model should be developed.

4.1. Computation of Mössbauer Spectra from a Theoretical Model

An excellent program to calculate theoretical spectra has been written by Gabriel and Ruby [28]. Starting with the Hamiltonian for the quadrupole and magnetic dipole interactions, it is necessary to specify the intensity of the magnetic field and the two angles that give its direction, the size of the electric field gradient (EFG), its asymmetry parameter, and the three orientation angles for each nuclear state. With these eight parameters the energy levels of the nuclear hyperfine splitting and their eigenvectors can be computed. The program is subdivided into three main parts: (a) computation of a single crystal Mössbauer spectrum, (b) of a powdered magnetic material in the absence of an external field, and (c) of a powdered material in an external magnetic field.

4.2. Curve Fitting of Mössbauer Data by Least-Square Analysis

This type program fits a series of Mössbauer resonance lines superimposed upon a parabolic baseline by the method of minimizing the value of the sum-squared residuals of the deviations. Since the Lorentzian profile used is not a linear function, it is linearized by approximations and the technique of successive iterations is used. The Lorentzian profile can be written as

$$Y(x, A, h, p) = \frac{A}{1 + (x - p)^2/B^2} = \frac{A}{1 + h(x - p)^2} \tag{7}$$

where A is the amplitude of the peak located in channel $x = p$, with a full width at half maximum of $2B = 2/\sqrt{h}$. To linearize this function, the following substitutions can be made, with H and P for the initial estimates:

$$h = H + \delta \qquad \delta \ll H$$
$$p = P + \gamma \qquad \gamma \ll P$$

Equation (7) expanded in a Taylor series results in

$$Y(x, A, C, D) = \frac{A}{Q} + \frac{C(x - P)^2}{Q^2} + \frac{D(x - P)^2}{Q^2} \tag{8}$$

letting $C = 2AH\gamma$, $D = -A\delta$, and $Q = 1 + H(x - P)^2$.

Equation (8) is linear and can be used in the least-square analysis, from which the parameters A, C, and D are obtained, and values for δ and γ can be calculated. Added to the initial estimates of half width and position, these new estimates are used as input in the iteration process. Details of the mathematics and logic of this type program have been described by Rhodes et al. [29], and a listing of the program is given in [7]. A time-sharing program is also available [30].

4.3. Curve Fitting of Mössbauer Data by Constrained Least-Square Analysis

In many cases the Mössbauer spectrum is too complex (a magnetic material with several lattice sites or a mixture with many components) for the programs described in Section 4.2. Since each peak requires three parameters and three are required for the baseline, the number of parameters is $3N + 3$, where N is the number of peaks. This requires the solution of $3N + 3$ simultaneous equations for each iteration, which for a large number of peaks drastically increases the demand of computer core storage and running time. The constrained program now makes use of the theoretical relationship that exists between the parameters of the Mössbauer lines. Thus, a magnetic iron spectrum could be fitted with five parameters (amplitude, centroid position, half width, quadrupole splitting, and internal magnetic field) instead of the 18 parameters required in an unconstrained program. Such a program has been published by Chrisman and Tumolillo [31]. The optimum application of computers in Mössbauer spectroscopy would be to couple the theoretical calculations described in Section 4.1. as an input for the curve-fitting model for the constrained program. At present this is not economically feasiable for the medium-size computer.

REFERENCES

1. H. Frauenfelder, *The Mössbauer Effect* (W. A. Benjamin, New York, 1962), Chap. 3.
2. E. Kankeleit, *Rev. Sci. Instr.* **35**, 194 (1964).
3. F. C. Ruegg, J. J. Spijkerman, and J. R. DeVoe, *Rev. Sci. Instr.* **36**, 356 (1965).
4. J. R. DeVoe, Ed., *NBS Tech. Note No.* **248** (1964), p. 29.
5. G. K. Wertheim, *Physics Today* **20**, 31 (1967).
6. R. Riesenman, J. Steger, and E. Kostiner, *Nucl. Instr. Methods* **72**, 109 (1969).
7. *NBS Misc. Publ. No.* **260-13** (1967).
8. J. R. DeVoe, Ed., *NBS Tech. Note No.* **276** (1966), p. 84.
9. R. Fritz and D. Schulze, *Nucl. Instr. Methods* **62**, 317 (1963).
10. P. A. Flinn, *Rev. Sci. Instr.* **34**, 1422 (1963).
11. C. Protop and C. Nistor, *Rev. Roum. Phys.* **12**, 653 (1967).
12. G. K. Wertheim, *Mössbauer Effect* (Academic Press, New York, 1964), Chap. 4.
13. M. W. Holm, Ed., Debye Characteristic Temperatures Table and Bibliography, *U.S. At. Energy Comm. Rept. No.* **ID-16399** (1957).
14. U. Shimony, *Nucl. Instr. Methods* **37**, 350 (1965).
15. L. May and D. K. Snediker, *Nucl. Instr. Methods* **55**, 183 (1967).
16. R. M. Housley, N. E. Erickson, and J. G. Dash, *Nucl. Instr. Methods* **27**, 29 (1964).
17. R. M. Housley, *Nucl. Instr. Methods* **35**, 77 (1965).
18. M. Kalvius, *Mössbauer Effect Methodology* **1**, 163 (1965).
19. N. Benczer-Koller and R. H. Herber, in *Chemical Applications of Mössbauer Spectroscopy*, V. I. Gol'danskii and R. H. Herber, Eds. (Academic Press, New York, 1968), p. 114.
20. J. C. Travis and J. J. Spijkerman, *Mössbauer Effect Methodology* **4**, 237 (1968).
21. W. A. Steyert and M. D. Daybell, *Mössbauer Effect Methodology* **4**, 3 (1968).
22. F. van der Woude and G. Boom, *Rev. Sci. Instr.* **36**, 800 (1965).
23. B. Sharon and D. Treves, *Rev. Sci. Instr.* **37**, 1252 (1966).
24. J. R. DeVoe, Ed., *NBS Tech. Note No.* **501** (1969), p. 7.
25. K. R. Swanson and J. J. Spijkerman, *J. Appl. Phys.* **41**, 3155 (1970).
26. J. H. Terrell and J. J. Spijkerman, *Appl. Phys. Letters* **13**, 11 (1968).
27. J. R. DeVoe, Ed., *NBS Tech. Note No.* **501** (1969), p. 17.
28. J. R. Gabriel and S. L. Ruby, *Nucl. Instr. Methods* **36**, 23 (1965).
29. E. Rhodes, A. Polinger, J. J. Spijkerman, and B. W. Christ, *Trans. Met. Soc. AIME* **242**, 1922 (1968).
30. L. May, S. J. Druck, and Martha Sellers, *U.S. At. Energy Comm. Rept. No.* **NYO-3798-2** (1968).
31. B. L. Chrisman and T. A. Tumolillo, *Computer Analysis of Mössbauer Spectra*, Dept. of Physics, Univ. of Illinois, Urbana, Ill. Available from the Clearing House, U.S. Dept. of Commerce, Springfield, Va. 22151, as Document AD-654 929 (1967).

Chapter 3

Nuclear Properties Determined from Mössbauer Measurements[1]

David W. Hafemeister

Physics Department
California State Polytechnic College
San Luis Obispo, California

In the short decade that has passed since Mössbauer's discovery, the Mössbauer technique has steadily shifted from the arena of the nuclear physicist to that of the solid-state physicist, chemist, or biologist. This is evidenced by the fact that many of today's leaders in Mössbauer spectroscopy are physicists who used to be concerned with the measurement of nuclear spins, parities, and beta-decay ft values, and who are now concerned with measurements of electron density distributions, relaxation and diffusion times, and magnetism. It is this very dual nature of the Mössbauer effect that makes it the vital topic that it is; the opportunity to be involved in both the nuclear and solid-state disciplines. We note that many of the measurements evaluate an observable **O** which is the product of a nuclear factor **N** and an atomic factor **A**:

$$\mathbf{O} = \mathbf{N} \cdot \mathbf{A}$$

Since in most cases the atomic factor **A** can be more readily calculated than the nuclear factor **N**, we will rely on our knowledge of the solid state to extract the nuclear information.

The present predominance of the solid-state discipline in Mössbauer studies does not mean that there are not any significant nuclear problems to attack, there are only fewer of them. It is the purpose of this chapter to discuss some of the following complex properties of nuclei that have been explored by Mössbauer measurements:

Section 2. Differences in the nuclear charge radius $\Delta R/R$.

[1] Supported by the National Science Foundation and the Office of Naval Research.

In order to set the stage for these topics it will be useful to expand on the classical analogies that Professors Debrunner and Frauenfelder set down in the first chapter. Classical pictorial analogies of the hyperfine interaction for the cannon shell will be discussed in Section 1.

1.   A PHENOMENOLOGICAL VIEW OF THE HYPERFINE INTERACTIONS

It is not our intention to go into the hyperfine interactions in great depth since in succeeding chapters Dr. Travis will discuss in detail the electric quadrupole interaction (Chapter 4) and Dr. Gonser will discuss in detail the magnetic interaction (Chapter 8). The beautiful example described in Chapter 1 of the "recoilless-floating-frozen-in-place-cannon" can be extended by analogy to understand the *small* hyperfine effects. We must remember that analogies are meant to give a physical understanding to a complex phenomenon and they should not be carried too far. Figures 1, 2, and 3 are pictorial descriptions of what happens to our cannon shell when we

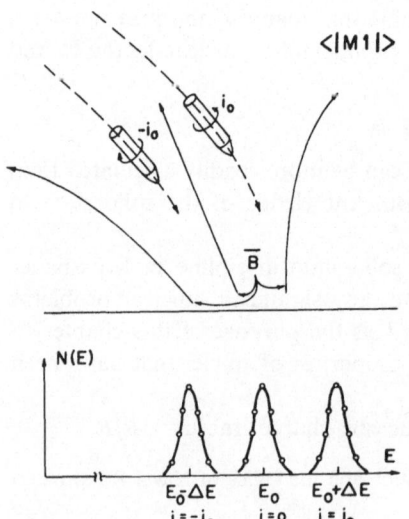

Figure 1.   The interaction of the magnetic dipole moment of a cannon shell with the earth's magnetic field. The range (energy) of the cannon shell is dependent on the orientation of the dipole moment.

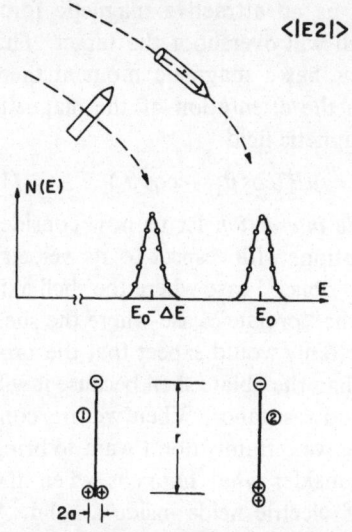

Figure 2.   The orientation of the quadrupole moment of a cannon shell. Wind resistance causes the "oblate" cannon shell to fall short of its mark.

consider that the shell has internal properties which will interact with its environment.

For the case of the *magnetic dipole hyperfine interaction,* let us consider that the cannon shell has a battery, some wire, and a current reversing switch. Consider the interaction between the current in the loop and the diverging lines of the earth's magnetic field B (see Figure 1). When the current flows in a clockwise ($i = -i_0$) direction, the component of the B field normal to the cannon shell will exert a repulsive force [$F = (\mu \cdot \nabla)B$] on the current loop and hence on the cannon shell. This will cause the cannon shell to fall slightly short of its former target. If now we reverse the current on the next cannon shell so that it flows in the counterclockwise ($i = i_0$) direction,

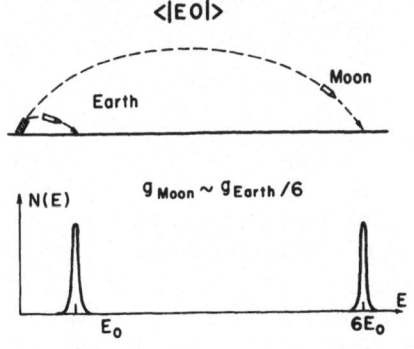

Figure 3.   The mass–mass monopole interaction. The range of the cannon shell is increased by reducing the force constant.

the earth's magnetic field will now give us an attractive magnetic force on the current loop. Now the cannon shell will overshoot the target. This analogy merely tells us that if the nucleus has a magnetic moment there will be a net gain or loss of energy when the orientation of the magnetic moment μ is changed with respect to a magnetic field.

$$\Delta E_{21} = (-\,\mu \cdot \mathbf{B})_2 - (-\,\mu \cdot \mathbf{B})_1 = \mu B \,(\cos \theta_1 - \cos \theta_2) \tag{1}$$

For the case of the *electric quadrupole interaction* let us now consider that the cannon shell can have two orientations with respect to its velocity vector (see Figure 2). We can think of the "oblate" case where the shell axis is perpendicular to its velocity vector and the "prolate" case where the shell axis is parallel to its velocity vector. We certainly would expect that the prolate shell will have a slightly longer range than the oblate shell because it will be more streamlined and will have less wind resistance. When we are considering the *actual* nuclear quadrupole case, we certainly don't want to bring in wind resistance, but we do want to consider what happens when the quadrupole moment is in the presence of electric fields inside a solid. A simple electrostatic problem will convince you that the energy of a two-proton nucleus will depend upon the angle of orientation of its quadrupole moment with respect to an external electron. When the axis of the nucleus shown in the lower part of Figure 2 is perpendicular to the electron–nucleus axis, the potential energy of the system is given by

$$W_1 = \frac{-\,2e^2}{\sqrt{a^2 + r^2}} \simeq \frac{-\,2e^2}{r} + \frac{e^2 Q}{2r^3} \tag{2}$$

where $Q = 2a^2$. If now the nucleus is rotated by $90°$ so that its axis is parallel to the electron–nucleus axis, its potential energy is reduced

$$W_2 = -\,e \left(\frac{e}{r + a} + \frac{e}{r - a} \right) \simeq -\,\frac{2e^2}{r} - \frac{e^2 Q}{r^3} \tag{3}$$

The energy of rotation is

$$\Delta W_{21} = W_2 - W_1 = -\,\frac{3e^2 Q}{2r^3} = +\,0.75 \, eqQ \tag{4}$$

where

$$q = V_{zz} = \frac{\partial^2 V}{\partial z^2} = -\,2e/r^3 \tag{5}$$

is the electric field gradient and eqQ is the quadrupole coupling constant.

Since the *electric monopole interaction* is caused by the interaction of the nuclear charge with the electrostatic field of the atomic electrons, let us consider the *gravitational monopole interaction* of the cannon shell with the gravitational field of the earth (see Figure 3). If we were to shoot our cannon shell on the moon, its range ($R \propto g^{-1}$) would be about six times greater than its range on earth (R_0) since g_{moon} is about 1/6 g_{earth}. By varying the strength of the mass–mass interaction we have changed the range of the projectile, and in a like manner by changing the strength of charge–charge interaction we will change the energy of the nuclear gamma ray. In actuality the nucleus has different charge radii for its ground and excited states. If the charge density in the source and absorber are unequal, the amount of energy given up in compressing the nucleus in the source will not be equal to the amount of energy gained in expanding the nucleus in the absorber. Thus, if the electrostatic spring constants in the source and absorber are unequal, the energy of the Mössbauer resonance will be shifted.

Now let us more formally consider the isomer shifts: the energy levels of a "finite-sized" nucleus of constant charge density will be slightly shifted from their value for the "point-sized" nucleus by the electrostatic interaction between the nucleus and its electronic cloud. The potential inside the nucleus of radius R is simply given by

$$V(r < R) = \frac{Ze^2}{R}(-3/2 + r^2/2R^2) \tag{6}$$

where the potential for the point nucleus is $V(r) = -Ze^2/r$. By integrating these two potentials over the nuclear volume and taking the difference, we obtain

$$\Delta E = \tfrac{2}{5}\pi Ze^2 R^2 |\psi_s(0)|^2 \tag{7}$$

The electronic density at the nucleus $|\psi_s(0)|^2$ will approximately be a constant over the nucleus and it is mainly from the s electrons. This interaction raises the nuclear levels because the spreading out of the nuclear charge dilutes the attractive electron–nucleon electrostatic interaction. It follows that the isomer shift δ is the difference in the source S and absorber A transition energies and is given by

$$\delta = \frac{c}{E_\gamma}[(E_{ex}^A - E_{gnd}^A) - (E_{ex}^S - E_{gnd}^S)] \tag{8}$$

$$\delta = \frac{2\pi c Ze^2}{5E_\gamma}(R_{ex}^2 - R_{gnd}^2)\left(\sum_A \psi^2(0) - \sum_S \psi^2(0)\right) \tag{9}$$

or

$$\delta = \frac{4\pi c Ze^2 R^2}{5E_\gamma}\left(\frac{\Delta R}{R}\right)\left(\sum_A \psi^2(0) - \sum_S \psi^2(0)\right) \tag{10}$$

where $\Delta R = R_{ex} - R_{gnd}$. The electron density near the nucleus is considerably increased by relativistic effects. These relativistic enhancement factors $S(Z)$ have been tabulated by Shirley [1], and they are in reasonable agreement

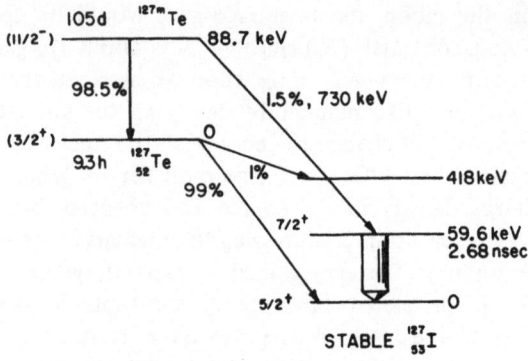

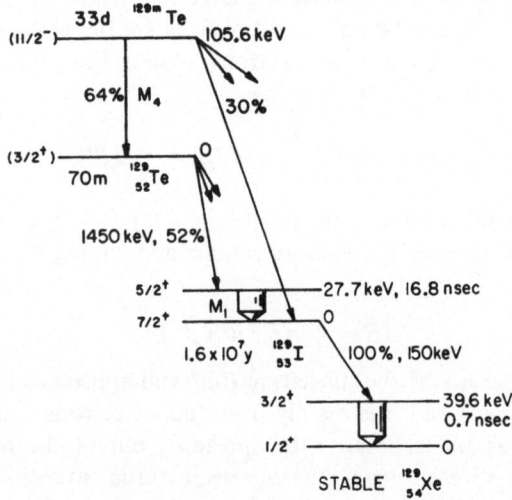

Figure 4. The relevant parts of the ^{127}I, ^{129}I–^{129}Xe decay schemes. The Mössbauer transitions are indicated by the three-dimensional arrows.

with a compilation of relativistic Hartree–Fock wave functions [2]. For iodine this factor is 2.8 and for uranium it is 18.2.

In order to give more variety to our discussion of these hyperfine effects we will not confine our discussion to the usual nuclear spin (3/2 to 1/2), as in metallic nuclei like ^{57}Fe, ^{119}Sn, or ^{197}Au, but we will discuss the higher spin (5/2 to 7/2), as in nonmetallic nuclei ^{127}I and ^{129}I. In Figure 4 the Mössbauer transitions are indicated by the three-dimensional arrows. ^{127}I is the 100% naturally-occurring isotope of iodine, whereas ^{129}I is the long-lived (1.6 × 10^7 yr) fission product. Each of these isotopes has its own merits, which must be considered when performing iodine experiments.

2. DIFFERENCES IN THE NUCLEAR CHARGE RADIUS $\Delta R/R$

The differences in the nuclear charge radius can be measured by a variety of techniques. The isotope shift, which is the difference between ground-state radii of different isotopes of the same element $[\Delta R/R = R(Z_1, A_2) - R(Z_1, A_1)]$, can be obtained from an analysis of μ-mesic x rays, K x rays, and optical data. The isotone shift which is the difference between the ground-state radii of nuclei that have the same neutron number $[\Delta R = R(Z_2, A_2, N_1) - R(Z_1, A_1, N_1)]$ is currently being measured by the μ-mesic x-ray technique. In this section we will concentrate on a description of the isomer shift, the change of nuclear radius between an excited state and the ground state for the same nucleus $[\Delta R = R_{ex}(Z_1, A_1) - R_{gnd}(Z_1, A_1)]$. The isomer shift can be measured by both the Mössbauer effect and by μ-mesic x-ray techniques. Thus far the μ-mesic data have mostly been obtained for the even–even rotational nuclei, but where the data exist for both techniques they are in fairly good agreement [3].

In order to obtain $\Delta R/R$ from a Mössbauer experiment it is necessary to be able to calculate [Eq. (10)] the difference in the electronic density at the nucleus for two compounds $\Delta|\psi(0)|^2 = |\psi_2(0)|^2 - |\psi_1(0)|^2$. It is at this point that the complications enter. The use of the free–ion estimates for $|\psi(0)|^2$ neglects the environment of the ion in the lattice. Corrections to $|\psi(0)|^2$ for the perturbations of the solid–state environment have been attempted for overlap, dipole–dipole polarization, hybridization, covalency, shielding, and band-theory interactions, but not always with great certainty. At present the calculation of the value of $\Delta R/R$ for ^{57}Fe is at variance by a factor of three [4], and at one time the sign of $\Delta R/R$ for ^{119}Sn was in doubt. These discrepancies do not mean that all values of $\Delta R/R$ have such large errors for there are many agreements and cross checks which indicate that $\Delta R/R$ can be obtained to within 20%. Recently Shenoy and Ruby [5] have reanalyzed much of the previous data using the properties of an isoelectronic series with the result that they have removed many of the ambiguities. In the paragraphs below we will discuss the various techniques that have been used to obtain $\Delta R/R$ from Mössbauer data.

2.1. $\Delta|\phi(0)|^2$ From "Pure Ionic" Valence States

The first calculation of the isomeric nuclear volume effect was carried out for ^{57}Fe, where $|\psi(0)|^2$ was obtained from the difference between the ionic ferrous and ionic ferric valence states. The change of $\Delta|\psi(0)|^2$ for this case is caused primarily by the change in the number of the $3d$ electrons. Since ferric iron ($3d^5$) has fewer d electrons than ferrous iron ($3d^6$), the effective nuclear charge for the ferric $3s$ electrons will be larger than in the ferrous case with the result that $|\psi_{3s}{}^{3+}(0)|^2 > |\psi_{3s}{}^{2+}(0)|^2$. Using the experimental data $\delta(Fe^{2+}) - \delta(Fe^{3+})=0.90 \pm 0.03$ mm/sec and the Hartree–Fock free ion wave functions, Walker et al. [6] obtained $\Delta R/R = -18 \times 10^{-4}$.

2.2. $\Delta|\phi(0)|^2$ From Band Theory

An analysis of the isomer shift for iron metal is useful since it serves as an additional calibration point on the ^{57}Fe isomer-shift scale. Most band-theory calculations usually present their results in terms of density-of-states curves and diagrams displaying the relationship between energy and wave vectors for the various bands and not in terms of electron densities. Ingalls [7] has extended the conventional band calculations by extracting information about the charge density. In order to do this he has decomposed the density-of-states curve that was obtained from band theory into the effective numbers of $3d$, $4s$, and $4p$ electrons. These populations were then used to calculate the direct contribution to $|\psi(0)|^2$ from the $4s$ electrons as well as the effect of shielding of $3s$ electrons by the $3d$ electrons. These techniques were then applied to the bcc phase, to the pressure dependence of the isomer shift in the bcc phase, to the bcc–fcc phase transition, and to the bcc–hcp phase transition with the result that $\Delta R/R \simeq -14 \times 10^{-4}$. Ingalls concludes that such band-theory calculations cannot accurately determine $\Delta R/R$.

2.3. $\Delta|\phi(0)|^2$ Caused by Overlap

As a result of the Pauli exclusion principle the free-ion wave functions are distorted by the overlap mechanism. To explain the macroscopic properties of the alkali halides, Löwdin [8] has introduced the symmetrical orthogonalization technique. He has shown that an atomic orbital ψ_μ in an ionic crystal can be given by

$$\psi_\mu = \sum_\alpha \phi_\alpha (1+S)_{\alpha\mu}^{-1/2}$$
$$= \phi_\mu - \tfrac{1}{2} \sum_\alpha \phi_\alpha S_{\alpha\mu} + \tfrac{3}{8} \sum_{\alpha,\beta} \phi_\alpha S_{\alpha\beta} S_{\beta\mu} \cdots \qquad (11)$$

with the summation extending over all neighbors in the crystal; the ϕ_α satisfy the free-ion Hartree–Fock equations, and $S_{\alpha\mu}$ is the overlap matrix expressed in the free-ion Hartree–Fock basis, and is given by

$$S_{\alpha\mu} = <\phi_\alpha|\phi_\mu> - \delta_{\alpha\mu} \qquad (12)$$

where $\delta_{\alpha\mu}$ is the Kronecker delta. Thus, the ψ_μ in Eq. (11) is an orthonormal set of atomic orbitals which use the Hartree–Fock free-ion basis set and acknowledge the nonzero overlap between the neighbors in the crystal. Thus far the deformed atomic orbitals ψ_μ have been applied to the isomer shifts of ^{129}I in the alkali iodides [9], and to the pressure dependence of the ^{57}Fe isomer shift in KFeF$_3$ [10]. In the case of ^{129}I, the change in the electron density due to overlap $[\Delta|\psi(0)|^2 \, \alpha \sum_{\mu\nu}(S_{\mu\nu})^2]$ for the alkali iodide series has the

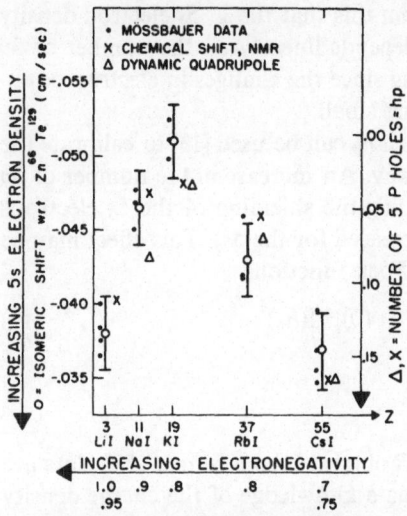

Figure 5. Isomer shifts, and the number of iodine p holes obtained from NMR chemical shift and from dynamic quadrupole coupling vs alkali atomic number for the alkali iodides from Refs. [11–13].

same trend [9] as the isomer-shift data shown in Figure 5. These trends are a result of the *competing* nearest neighbor alkali–iodide and next nearest neighbor iodide–iodide overlaps. When the alkali Z is increased, the alklali-iodide overlaps increase because of the increased alkali size while the iodide-iodide overlaps decrease because their separation with distance increases. LiI and CsI have the most overlap because of the I–I and Cs–I overlaps, respectively.

For the case of ^{57}Fe, the value $\Delta R/R < -5.2 \times 10^{-4}$ was established [10] as an upper limit. Since covalency increases with pressure it was not possible to determine the actual value, but this limiting value is a factor of three smaller than the values quoted in [6] and [7].

2.4. $\Delta|\phi(0)|^2$ From the Shielding of p Holes

Data on the number of iodine $5p$ electrons for the alkali iodides are available from the NMR chemical shift measurements of Bloembergen and Sorokin [11] and from the dynamic quadrupole-coupling measurements of Menes and Bolef [12]. In the NMR method, the change of the magnetic field at the nucleus caused by the paramagnetism of the $5p$ electrons is measured. In the dynamic quadrupole-coupling method, the attenuation of a sound wave caused by the nuclear spin–phonon interaction is measured. Both of these effects are proportional to the number of $5p$ holes, $h_p = 6 - y$, where the I$^-$ configuration may be written as $5s^2 5p^y$. The xenon configuration corresponds to $h_p = 0$ and $y = 6$. The values of h_p determined from these techniques are compared with the isomer shifts of the alkali iodides in Figure 5. The behavior of h_p is similar to that of the isomer shifts, and as a first approximation, we can assume from this that the I^- $5s$ electron density as measured by the Mössbauer effect depends linearly on the number of $5p$ holes, h_p. This linearity is not surprising since the changes in electron populations, $h_p \sim 0.1$, in the alkali iodides, are small.

The linear dependence between δ and h_p can be used [13] to calibrate the ^{129}I isomer-shift scale in the following way. An increase in the number of $5p$ electrons will decrease $|\psi(0)|^2$ by increasing the shielding of the $5s$ electrons (the effective nuclear charge will be decreased for the $5s$). This effect may be expressed quantitatively with Hartree–Fock functions

$$\Delta|\psi(0)|^2 = 0.058|\psi(0)|^2\Delta(h_p) \tag{13}$$

to obtain $\Delta R/R$.

2.5. ΔR Ratio Method

Discrepancies in the values of $\Delta R/R$ obtained from isomer-shift data are directly traceable to the present inadequate knowledge of the charge density of an ion when it is situated in a lattice. It is possible to overcome part of this difficulty by making ratios of the ΔR values from the isomer-shift data obtained from two different Mössbauer levels

$$\frac{(\Delta R/R)_2}{(\Delta R/R)_1} = \frac{E_{r_2}}{E_{r_1}} \frac{\delta_2}{\delta_1} \frac{Z_1 R_1^2}{Z_2 R_2^2} \frac{S(Z_1)}{S(Z_2)} \frac{\Delta\psi^2(Z_1)}{\Delta\psi^2(Z_2)} \tag{14}$$

For the case of two excited states in the *same* nucleus (e.g., the 97.4- and 103.2-keV levels in ^{153}Eu), this expression becomes

$$\frac{(\Delta R/R)_2}{(\Delta R/R)_1} = \frac{E_{r_2}}{E_{r_1}} \frac{\delta_2}{\delta_1} \tag{15}$$

For the case of two different *isotopes* of the same element (e.g., the 57.5-keV level in ^{127}I and the 27.7-keV level in ^{129}I) it becomes

$$\frac{(\Delta R/R)_2}{(\Delta R/R)_1} = \frac{E_{r_2}}{E_{r_1}} \frac{\delta_2}{\delta_1} \frac{R_1^2}{R_2^2} \tag{16}$$

Thus far about a dozen ratios have been obtained from these two approaches.

This method has been extended to isoelectronic compounds by Shenoy and Ruby [5] who have obtained the ratios of nuclear radii for nuclei from tin, $Z=50$, to xenon, $Z=54$. By considering the isomer shifts of such isoelectronic pairs as $(^{129}IO_4)^-$ vs $^{129}XeO_4$, $^{129}I^-$ vs $^{129}Xe^\circ$, and $K_2^{125}TeO_4$ vs $K^{129}IO_4$, they found that the isomer shifts for the neighboring Mössbauer isotopes were linearly related. (See Figure 6 for a comparison of the ^{125}Te and ^{129}I shifts.) This linearity implies that the bonding in the neighboring isoelectronic compounds is indeed similar. Their calculations with HF–

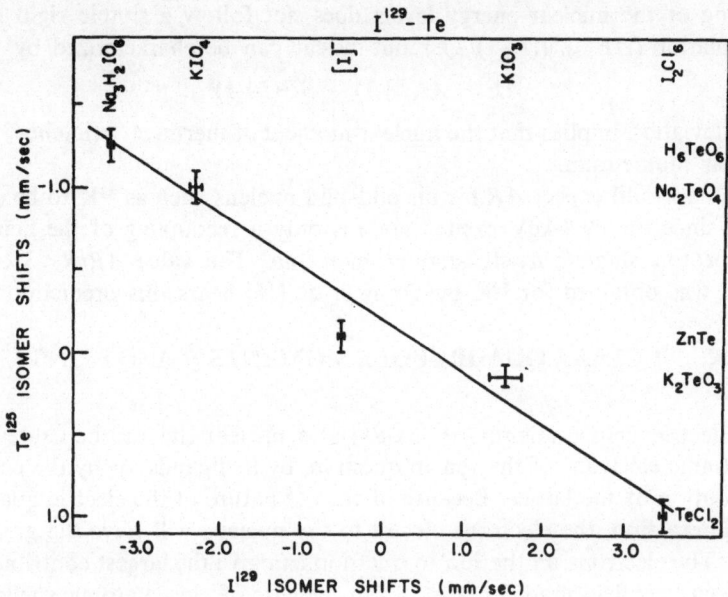

Figure 6. Plot of isomer shifts of the isoelectronic pairs of compounds of [125]Te and [129]I taken from the work of Shenoy and Ruby [5].

SCF wave functions showed that the ratio of electron densities for any pair of isoelectronic compounds was approximately a constant $\Delta\psi^2(Z+1)/\Delta\psi^2(Z) \simeq 5/4$ for the various $5s^m5p^n$ configurations. The values of ΔR which these authors obtained from the isoelectronic ratio method are in quite good agreement with their direct density calibrations obtained with Eq. (10).

2.6. Nuclear Information from ΔR

Certainly the growing interest in measurements and calculations of the monopole moment differences puts $\Delta R/R$ on an equal footing with the more traditional nuclear moments, the magnetic dipole moment and the electric quadrupole moment. As in the case of μ and Q, the agreement between theory and experiment is not yet impressive. The pairing plus quadrupole calculations of Uher and Sorensen [14] for odd A spherical nuclei are only in fair agreement with the data. In a similar fashion for the even–even rotational nuclei Marshalek's [15] calculations which employ the centrifugal stretching and Coriolis antipairing interactions are also only in fair agreement. Some of the present data [3] shows that $\Delta R/R$ for the tungsten isotopes is small or even *negative*. One would expect ΔR to be *positive* since the spacing of the nuclear energy levels does not follow a simple rigid rotor formulation ($\Delta E = \hbar^2 I(I+1)/2\mathscr{I}$), but instead can be characterized by

$$\Delta E = AI(I+1) - BI^2(I+1)^2 \tag{17}$$

This deviation implies that the nuclear moment of inertia $\mathscr{I}$ "stretches" with angular momentum.

One would expect ΔR for an odd–odd nucleus such as ^{40}K to be quite small since the 29.4-keV excited state is only a recoupling of the neutron and proton single particle angular momenta. The value $\Delta R/R < 3 \times 10^{-4}$ which was obtained for ^{40}K by Tseng et al. [16] bears this prediction out.

3. NUCLEAR QUADRUPOLE MOMENTS Q AND SPINS I

Electric field gradients $\partial^2 V / \partial x_i \partial x_j$ at a nuclear site can be caused by the atomic electrons of the ion in question, by its ligands, or by the charge distribution of the lattice. Because of the r^{-3} nature of the electric quadrupole interaction, the electrons closest to the nucleus will have the greatest effect. The electrons on the ion in question can give the largest contribution to the electric field gradient (EFG). For the case of closed atomic shells the lattice contributions will predominate. The polarizable nature of the electron cloud enhances or degrades the EFG and it is described quantitatively by the Sternheimer shielding factors.

The quantum mechanical expression which describes the interaction between a nucleus with a quadrupole moment eQ and the EFG q is given by the Hamiltonian operator

$$H = \frac{eqQ}{4I(2I-1)}\left[3I_z^2 - I(I+1) + \frac{\eta}{2}(I_+^2 + I_-^2)\right] \tag{18}$$

where I_+ and I_- are the raising and lowering operators. The largest component of the EFG along a principal axis is given by $q = V_{zz} = \delta^2 V/\delta z^2$ and the asymmetry parameter is given by $\eta = (V_{xx} - V_{yy})/V_{zz}$, where the x and y axes are defined so that η is always less than one. For the case of $I = 3/2$ the shifts in the energy levels are obtained from

$$\Delta E = \frac{eqQ}{4I(2I-1)}[3m^2 - I(I+1)](1 + \eta^2/3)^{1/2} \tag{19}$$

When the crystal has axial symmetry, then $\eta = 0$ and the substates are characterized by definite m values with the $\pm m$ states being degenerate. The energy shifts can then be easily evaluated for all values of spin from the relationship

$$\Delta E = \frac{eqQ}{4I(2I-1)}[3m^2 - I(I+1)] \tag{20}$$

We will now describe how Mössbauer determinations of the nuclear quadrupole moment have been carried out for the following cases:

Section 3.1. Quadrupole moment ratios Q^*/Q.
  (i) ^{129}I ($I^*=5/2$, $I=7/2$) with $\eta=0$.
  (ii) ^{129}I with $\eta=-0.16$.
Section 3.2. Absolute quadrupole moment values.
  (i) Ferrous ^{57}Fe ($I^*=3/2$, $I=1/2$): valence EFG.
  (ii) Ferric ^{57}Fe: lattice EFG.

3.1. Quadrupole Moment Ratios Q^*/Q

Since the hyperfine lines are caused by the energy differences between the substates of the excited state and the ground state, a Mössbauer experiment will only determine the ratio Q^*/Q and not the values Q^* and Q separately. If the value of Q is known from other measurements, then Q^* is obtained from the ratio. The examples described below will discuss these points in detail for the case when the EFG is axially symmetric ($\eta=0$), and when it is not axially symmetric ($\eta\neq0$).

(i) ^{129}I ($I^*=5/2$, $I=7/2$ with $\eta=0$). The combination of a ZnTe source and an iodate absorber [13] gives a spectrum (Figure 7) that can be interpreted by pure quadrupole coupling in the iodate. Previous NMR measurements on the ^{129}I iodates have shown that the asymmetry para-

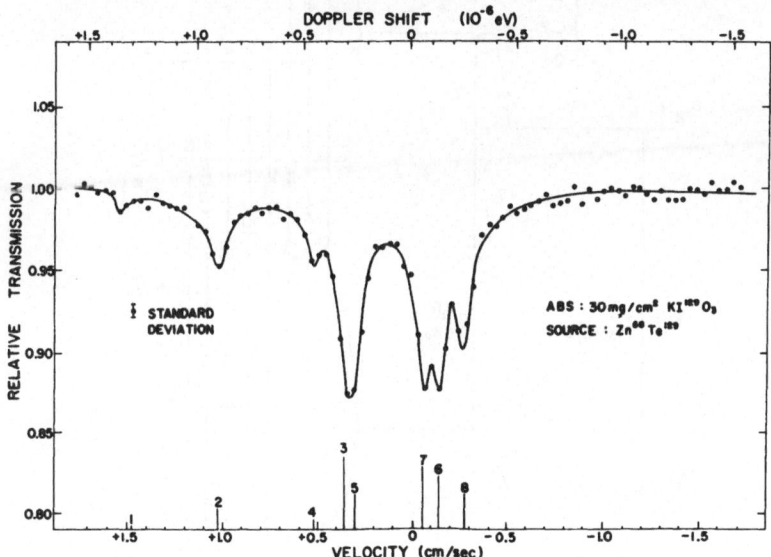

Figure 7. Quadrupole hyperfine structure of ^{129}I in KIO_3 with $\eta\simeq0$ from Hafemeister *et al.* [13].

meter η is very small and can be neglected. Hence, it is possible to represent the positions of the hyperfine absorption lines by

$$\delta_{ij} = \frac{ceqQ_{\text{gnd}}}{4E_r}\left[\frac{Q^*}{Q}C(I^*,m_j^*) - C(I,m_i)\right]+\delta' \qquad (21)$$

where

$$C(I,m) = [3m^2 - I(I + 1)]/I(2I - 1)$$

δ_{ij} is the shift of the transition from the ground-state level $|I, m_i>$ to the excited-state level $|I^*, m_j^*>$, and δ' is the isomer shift of the iodate with respect to the ZnTe source. The states considered here have spins $I^*=5/2$ and $I=7/2$, and the transition between them has pure $M1$ character (m_i-m_j $=0, \pm1$) so that there are eight components (see Figure 8). The $^{129}KIO_3$ ground-state coupling constant $eqQ_{\text{gnd}}=698.9$ Mc/sec was calculated from the known $^{127}KIO_3$ coupling constant of 996.7 Mc/sec measured at 80°K and the accurately measured quadrupole moment ratio $Q_{129}/Q_{127}=$

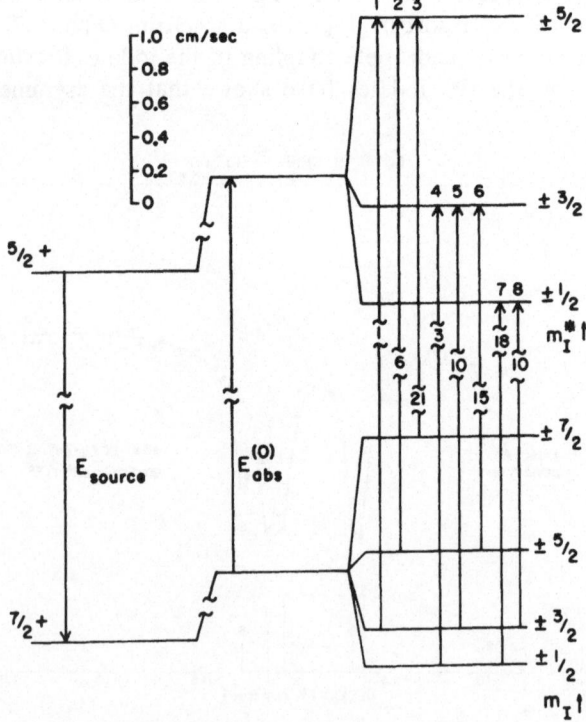

Figure 8.   Nuclear energy levels for ground and first excited states of ^{129}I in KIO_3.

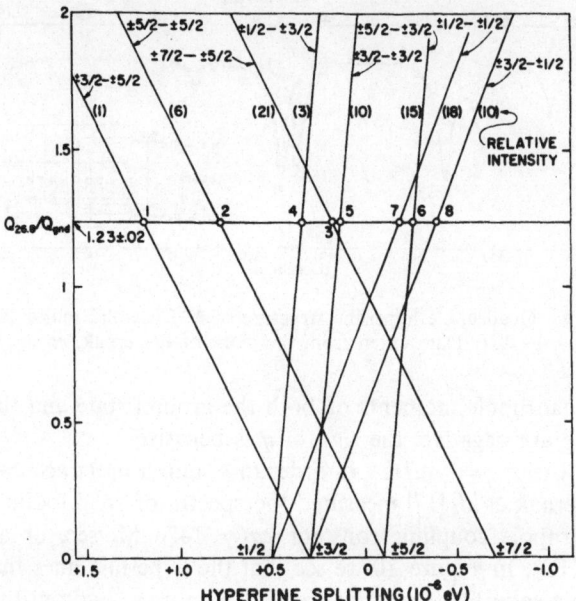

Figure 9. KIO_3 line positions computed from Eq. [21] as a function of the quadrupole ratio Q^*/Q.

+0.70121. Figure 9 is a representation of Eq. (21) and demonstrates how the line positions change as a function of Q^*/Q. If we compare Figure 9 with the measured spectrum of KIO_3 given in Figure 7, we see that the observed structure is obtained for $Q^*/Q = +1.23$. The line intensities, which for a thin polycrystalline absorber are proportional to the square of the Clebsch–Gordan coefficients $<I1m_i\Delta m \mid I1I^*m_j^*>$, are in agreement with the data.

The ratio $Q^*/Q = +1.23 \pm 0.02$ and the KIO_3 isomer shifts $\delta' = 0.156 \pm 0.02$ cm/sec were deduced from the measured values of δ_{ij} for lines 2, 4, 6, 7, and 8 by least squares fit to Eq. (21). This equation represents a straight line slope Q^*/Q and an intercept δ'. The KIO_3 quadrupole-coupling constants eqQ_{129}, which were determined from the spectra are in agreement with the more accurate NMR values for the [127]I iodates.

The sign of the field gradient q at the [129]I nucleus follows directly from the asymmetry of the hyperfine pattern (see Figures 7 and 8). This is true for nuclei in which at least one of the states has a spin larger than 3/2. By considering, for instance, transitions numbers 2 ($\pm 5/2 \rightarrow \pm 5/2$) and 6($\pm 5/2 \rightarrow \pm 3/2$), both starting from the same ground state, we see in Figure 7 that line 2 has higher energy than line 6. Thus the excited state with $m_j^* = \pm 5/2$ lies higher than $m_j^* = \pm 3/2$. This implies that eqQ is positive [see Eq. (21)].

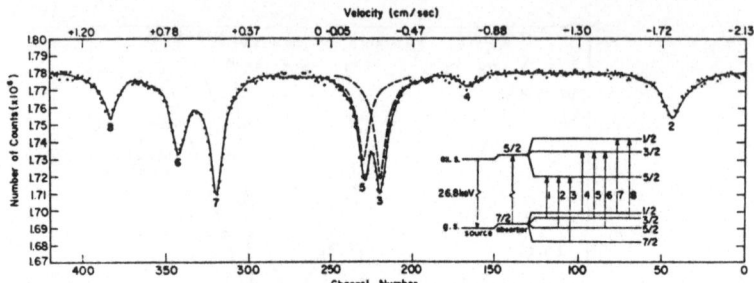

Figure 10. Quadrupole hyperfine structure of ^{129}I in solid molecular iodine
I_2 with $\eta = -0.16$. Data taken from the work of Pasternak, *et al.* [17].

Since the quadrupole moments of both the ground state and the 26.8-keV
excited state are negative, the sign of q is negative.

(ii) ^{129}I **with** $\eta = -0.16$.   In order to obtain a more accurate value of
Q^*/Q, Pasternak *et al.* [17] measured the spectra of solid iodine ^{129}I, which
has a quadrupole coupling constant $eqQ=1426$ Mc/sec, or about twice
that of $^{129}KIO_3$. In Figure 10 we see that the hyperfine lines that they obtained from a solid $^{129}I_2$ absorber are indeed more spread out than the corresponding lines for $^{129}KIO_3$. The analysis of the data is slightly complicated
by the fact that the asymmetry parameter is equal to -0.16. In order to take
the asymmetry parameter into account, these authors have used Bersohn's
calculation [18] for the energy spacings which have been expressed as an
expansion in terms of the even powers of η

$$\Delta E_i = A_i + B_i\eta^2 + C_i\eta^4 + D_i\eta^6 + E_i\eta^8 \tag{22}$$

By fitting the line positions to the differences in energy for the substates
of the ground state and excited state as represented by Eq. (22), Pasternak
et al. obtained $Q^*/Q=+1.232\pm0.004$ and $Q^*/Q=+1.237\pm0.002$. A comparison of Figure 7 for KIO_3 and Figure 10 for I_2 indicates that the two
spectra are approximately inverted from right to left. This inversion means
that the sign of eqQ is negative with the result that q is positive. Since the
field gradient at the iodine nucleus is due to a p electron vacancy which
acts like a positive charge, we expect a positive value for q (i.e., $V=+e/r$
and $\partial^2 V/\partial r^2 = +2e/r^3$).

3.2.   Direct Determination of Quadrupole Moment Values

If the spin of one of the nuclear states is 1/2 (as in the case of ^{57}Fe, ^{119}Sn,
and ^{197}Au), there will not be a quadrupole interaction in that state, and
we must then only consider the quadrupole interaction in the remaining state.
If the properties of the lattice and the ion are well known, then a calculation
of q can be performed so that the value of Q may be extracted from the

measured value of eqQ. In this section we will discuss the procedures that have been utilized for the determination of the quadrupole moment of the 3/2, 14.4-keV state of ^{57}Fe.

(i) Ferrous ^{57}Fe ($I^*=3/2$, $I=1/2$): valence EFG. An electric field gradient at an iron site will split the $I=3/2$, 14.4-keV level into 2 substates ($m=\pm 3/2$ and $\pm 1/2$), but will not effect the $I=1/2$ ground state. Two transitions are observed ($\pm 3/2 \rightarrow \pm 1/2$ and $\pm 1/2 \rightarrow \pm 1/2$) and from Eq. (19) the separation between the two resonances is

$$\Delta E = \frac{eqQ}{2}(1 + \eta^2/3)^{1/2} \tag{23}$$

Both the electrons on the ion as well as the lattice can contribute to the EFG

$$q = (1 - R)q_{\text{val}} + (1 - \gamma_\infty)q_{\text{lat}} \tag{24}$$

where $(1-R)$ and $(1-\gamma_\infty)$ are the Sternheimer shielding factors. The most important contribution to the EFG is from q_{val}, which is $(4/7)<1/r^3>_{3d}$ for the free ion in the ^5D configuration without the spin orbit interaction. However, Ingalls [19] has shown that the value of q_{val} will be reduced by about a factor of three if one considers the effects of the crystalline field at finite temperature, the spin orbit interaction, covalency, and Sternheimer shielding of $(1-R)=0.68$. By applying these considerations to the Mössbauer results in FeSiF$_6\cdot$6H$_2$O, Ingalls obtained for ^{57m}Fe $Q=0.29\pm 0.02$ b. Here again it is necessary for the nuclear physicist to rely on calculations of the solid state to extract a nuclear parameter. However, to cloud the issue somewhat these solid-state calculations are not universally accepted. For example, Nozik and Kaplan [20] have calculated Q_{lat} for these compounds, and they claim that the Q_{lat} contribution cannot be neglected. They claim that this will lower the value of Q to 0.20 b. Very recently, Chappert et al. [21] induced a field gradient on Fe^{2+} impurities in MgO by applying a strong magnetic field at low temperatures. From their analysis they obtained $Q=0.21\pm 0.03$ b. The nuclear theorists are not much help in these matters because at present they cannot reliably calculate nuclear quadrupole moments in nuclei as ^{57}Fe to better than a factor of two.

(ii) Ferric ^{57}Fe: lattice EFG. In order to determine the ^{57}Fe quadrupole moment from a somewhat different approach, Artman et al. [22] studied the quadrupole splitting of the ferric ion in $\alpha-$Fe$_2$O$_3$. Since this ion is in the ^6S state, no contribution is expected from q_{val}, and only q_{lat} must be considered. Sample data for α-Fe$_2$O$_3$ is shown in Figure 11, and we note that the six-line magnetic pattern predominates. For this case the quadrupole interaction only slightly asymmetrically shifts the purely symmetric magnetic pattern. Because $\mu H \gg eqQ$, the quadrupole-coupling constant can be

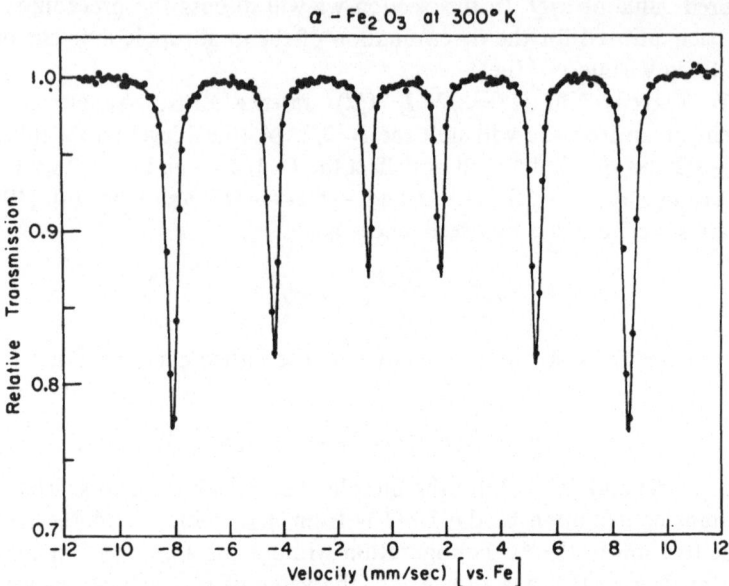

Figure 11. The magnetic dipole and electric quadrupole hyperfine structure of α-Fe$_2$O$_3$ taken from the work of Artman *et al.* [22]. The electric
quadrupole interaction is the cause of the asymmetry in the six-line magnetic pattern.

calculated directly from the slight shifts in the spectra. These authors have
then performed a lattice sum of the monopole and dipole contributions
to q_{lat} with the result that $Q = 0.283 \pm 0.035$ b.

3.3. Direct Determination of Nuclear Spins

Many determinations of the spins of excited nuclear states rely on
assumptions which are dependent on the details of a nuclear model. However, if the ground-state spin is already known, the Mössbauer hyperfine
spectra allows the spin of the excited state to be determined unambiguously.
Certainly the ^{57}Fe and ^{129}I quadrupole-coupling data can *only* be explained
if the spins of the excited states are 3/2 and 5/2, respectively. Thus far the
Mössbauer measurements of nuclear spin have not disagreed with any of
the prior model-dependent determinations, but their existence puts the
values of these spins on firmer ground.

4. NUCLEAR MAGNETIC DIPOLE MOMENTS

In an analogous fashion to the electric quadrupole interaction, the
magnetic hyperfine interaction can be used to determine values of the

nuclear magnetic dipole moment. The hyperfine Hamiltonian for a nuclear magnetic dipole μ in a magnetic field H is given by

$$H_M = - \mu \cdot \mathbf{H} = - g\mu_N \mathbf{I} \cdot \mathbf{H} \tag{25}$$

and the energy levels are obtained from

$$E_M = - g\mu_N H m_I \quad (m_I = I, I - 1, ..., -I) \tag{26}$$

where μ_N is the nuclear magneton and g is the gyromagnetic ratio. These equations indicate that the $2I+1$ magnetic sublevels are equally spaced with a separation of $g\mu_N H$ between the levels. The nucleus experiences this interaction in both its excited and ground states with the result that the energy of the gamma ray for a transition from the $m_i{}^*$ excited-state sublevel to the m_j ground-state sublevel will be

$$E_{ij} = E_0 + \frac{\delta}{c} E_0 - g\mu_N H m_j + g^* \mu_N H m_i{}^* \tag{27}$$

where the isomer-shift energy has been included. An NMR experiment measures the small energy differences between the magnetic substates of the ground state

$$\Delta E_{kj} = g\mu_N H (m_k - m_j) \tag{28}$$

As in the case of the electric quadrupole interaction, the measurement of the Mössbauer hyperfine splitting will yield a nuclear–atomic interaction strength $g\mu_N H$ and ratio of the gyromagnetic moments g^*/g. The equation for the individual Mössbauer hyperfine lines is given by

$$\Delta E_{ij} = \frac{\delta}{c} E_0 - g\mu_N H \left(m_j - \frac{g^*}{g} m_i{}^* \right) \tag{29}$$

which can be used to make a pattern of lines similar to the quadrupolar case shown in Figure 9. Note that a purely magnetic hyperfine interaction will give a symmetric pattern of absorption lines centered around its isomer shift. As in the case of the quadrupole interaction the Δm selection rules are determined by the multipolarity of the radiation, and the intensities of the transitions by the square of the appropriate Clebsch–Gordan coefficients. Unlike the quadrupolar case, the ground-state g factors are usually quite well known (to about one part in 10^4) because the interaction energy μH can be measured very accurately with NMR, and the paramagnetic and diamagnetic shielding factors are quite small ($\sim 0.01\%$) and much less than the Sternheimer shielding factors. By using the known ground-state value of μ, a measurement of the magnetic hyperfine splitting will give the excited-state magnetic moment, as well as the field strength H.

For the $2^+ \to 0^+$ even–even nuclei, the magnetic interaction in the ground state will not exist, and the hyperfine splitting will be entirely from the excited state. The magnetic measurements will then only give a value for $g^*\mu_N H$, and one must then know the value of g^* or H from another measurement in order to determine H or g^*.

In the remainder of this section, we will describe the determination of the magnetic moment of the 27.7-keV, $5/2^+$ state of ^{129}I. Since none of the iodide or telluride lattices have large magnetic fields, it is necessary to use of external magnetic field or to implant into an iron foil. In their first measurement of the magnetic moment of the $5/2^+$ state, deWaard and Heberle [23] used a single-line $Zn^{129}Te$ source and a $K^{129}I$ absorber in the field of a superconducting magnet. Since the external magnetic field was longitudinally applied along the direction of the gamma-ray beam, only

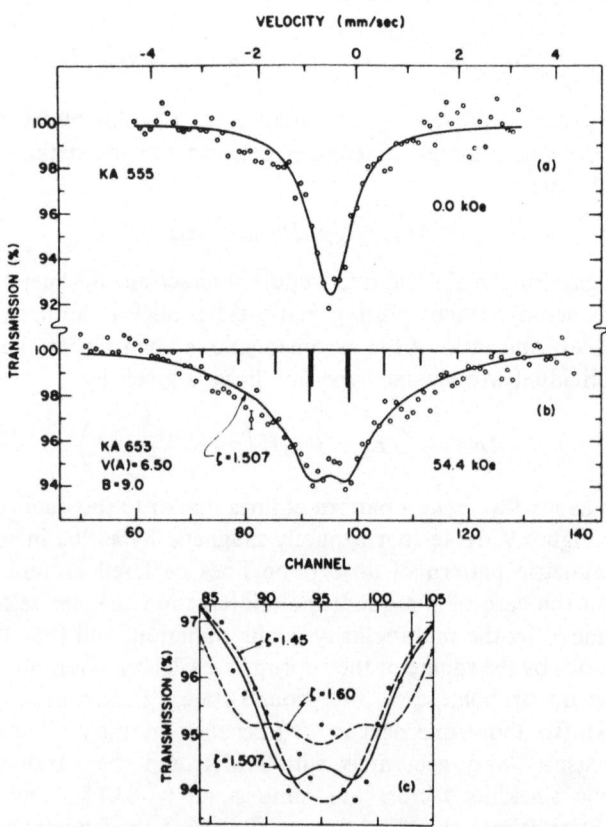

Figure 12.   Spectra of ^{129}KI in (a) zero magnetic field and (b) in a field of 54.4 kOe taken from the work of deWaard and Heberle [23]. The data has been fitted for several ratios of the Zeeman splittings.

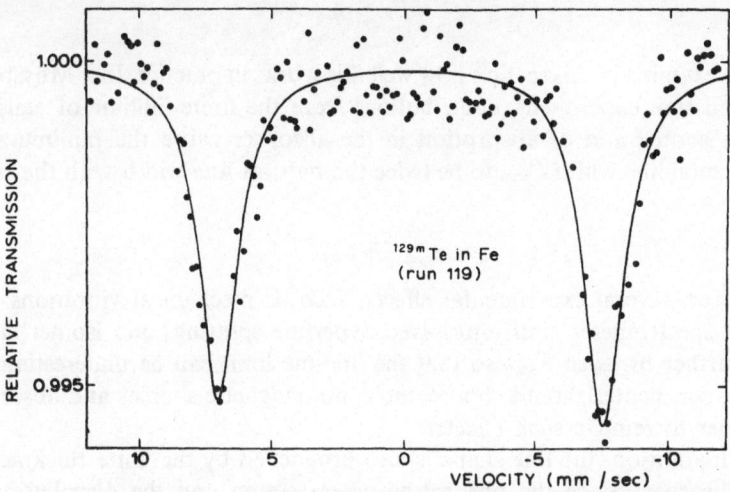

Figure 13. Mössbauer spectra of ^{129}I implanted as ^{129m}Te in an iron foil from deWaard; *et al.* [24].

the 12 $\Delta m = \pm 1$ transitions are absorbed. The spectra taken at 0 and 54.4 kOe (Figure 12) indicate that the line is considerably broadened by the magnetic field. By fitting the spectra to the sum of 12 Lorentzian curves, these authors obtained $\mu(5/2^+) = 2.84 \pm 0.05$ nm.

In a second measurement they [24] were able to obtain a much more dramatic splitting in the magnetic hyperfine structure. The 33-day isomer of ^{129}Te was ionized, accelerated, and then implanted into an iron foil. The conduction electron polarization effect, which has been discussed by Shirley and Westenberger [25], creates a massive (1130 ± 40 kOe) field at the substitutional iodine atoms. The lines in the spectra displayed in Figure 13 are cleanly separated.

Recently, Perlow *et al.* [26] observed a very large hyperfine anomaly of 7% in Mössbauer studies of ^{193}Ir. This anomaly appears as a change in the ratio g^*/g when spectra taken with internal and external magnetic fields are compared. This result indicates that the magnetic moment distributions for the two Mössbauer states of ^{193}Ir are quite different. In addition they have observed a 2% anomaly between ferromagnets and antiferromagnets.

5. NUCLEAR LIFETIMES

As was pointed out in the first chapter, the Mössbauer effect measures the natural line width Γ, which is related to the nuclear lifetime by the uncertainty principle

$$\Gamma\tau \geq \hbar \qquad (30)$$

Let us examine in this section how well this works in practice. In a Mössbauer transmission experiment, contributions from the finite lifetime of emission in the source and of absorption in the absorber cause the minimum experimental line width Γ_{exp} to be twice the natural line width with the result that

$$\tau \geq 2\hbar/\Gamma_{exp} \qquad (31)$$

However, several experimental effects, such as mechanical vibrations in a faulty spectrometer, and unresolved hyperfine splittings and isomer shifts, can further broaden Γ_{exp} so that the lifetime limit can be underestimated. The experimenter should choose cubic, nonmagnetic sources and absorbers in order to remove such effects.

In addition, the line shape is also broadened by the finite thickness of the absorber. Since the absorption cross section and the distribution of emitted gamma rays can both be described by a Lorentzian function $[(E-E_0)^2+\Gamma^2/4]^{-1}$, we see that the initial thickness of the absorber will remove more of the center part of the Lorentzian gamma-ray distribution than from its wings. Because of this, the gamma-ray line shape will be broader for the last part of the absorber than it was for the initial part. Figure 14 shows the calculated [27] thickness T, dependence of the line width Γ, the absorption amplitude I_A, and the area $A=0.5f\Gamma\pi L(t)$. Note

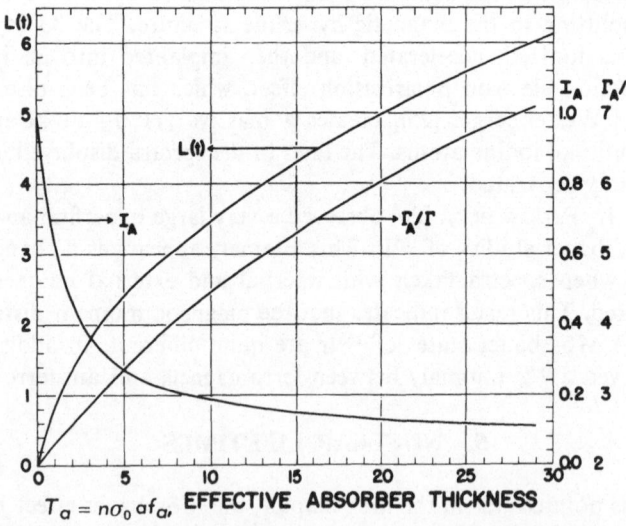

Figure 14. The Lorentzian function $L(t)$, the intensity I_A, and the relative line width Γ_A/Γ vs the effective absorber thickness T.

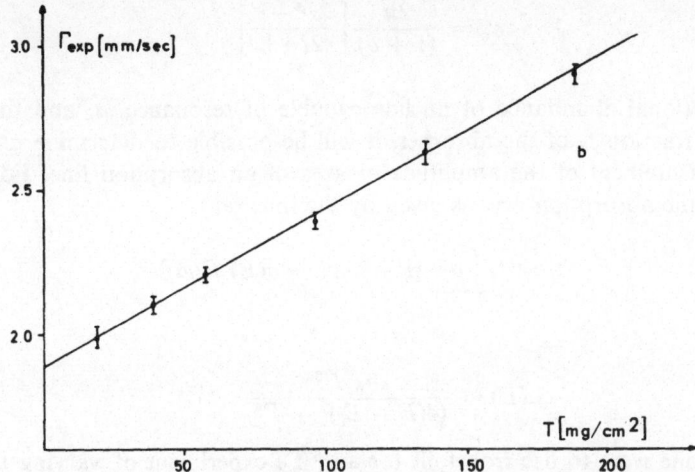

Figure 15. Line width of the [197]Au transmission spectra as a function of the thickness of metallic gold. The lifetime is obtained from the extrapolation to zero thickness from Steiner *et al.* [28].

that the amplitude saturates much more quickly than the area. The broadening effect is shown for the [197]Au data [28] in Figure 15. For small thicknesses of the absorber the line width increases linearly with the thickness. By fitting the data to the theoretical expressions for line broadening, the zero-thickness line width is obtained. In this particular case the Mössbauer measurement gave the result $\tau_{1/2}=2.730\pm0.020$ nsec. The quoted error on this experiment is less than 1% and is about as small as that obtained by the best electronic methods. Of course, the Mössbauer technique is only applicable for the range of lifetime of 10^{-7}–10^{-10} sec.

6. INTERNAL CONVERSION

Mössbauer techniques can also be applied to the measurement of the internal conversion coefficient, $a=N(e^-)/N(\gamma)$, the ratio of the atomic electron and nuclear gamma ray transition probabilities. In the early days of the Mössbauer effect, it was noted that the value of a needed to explain the amplitudes of the Mössbauer resonances was about half the published value of $a=15\pm1$, which had been obtained by conventional nuclear techniques. This discrepancy was resolved by the measurement of a by the Mössbauer effect.

Since the effective thickness $T_a=n\sigma_oaf_a$ of a Mössbauer absorber is the product of the number of atoms of element per cm² n, the resonance cross section

$$\sigma_0 = \frac{2\pi}{(1+a)}\left[\frac{2I^*+1}{2I+1}\right] \tag{32}$$

the fractional abundance of nuclide capable of resonance, a, and the recoilless fraction f_a of the absorber, it will be possible to determine a from the measurement of the amplitude or area of an absorption line. For example, the absorption area is given by the integral

$$A = f_s \int_{-\infty}^{\infty} dE \, [1 - \exp(-\sigma(E) \, f_a na)] \tag{33}$$

where

$$\sigma(E) = \frac{\sigma_0 \, \Gamma^2}{(4(E-E_0)^2 + \Gamma^2)} \tag{34}$$

If one were to use iron foils for an ^{57}Fe experiment of varying thickness in mg/cm^2, the amount of photoelectron absorption from the atomic electrons would vary from absorber to absorber giving incorrect values of the area. In order to remove this effect Hanna and Preston [29] used a series of iron absorbers, which all had the same thickness of iron atoms but whose density of ^{57}Fe atoms varied by as much as a factor of 37. By evaluation of the transmission integral (see Figure 14) and by measuring the ratios of the areas in order to remove f_s (recoilless fraction of the source), they obtained the value $a=8.9\pm0.7$, which agrees with the more recent nuclear coincidence measurements.

7. PARITY, MULTIPOLE MIXING, AND TIME REVERSAL

In a few cases the Mössbauer effect has been used to test symmetry principles. It is not our purpose to discuss these phenomena in depth, but merely to acquaint the reader slightly with these more esoteric topics. In Section 7.1. we shall discuss a β–γ Mössbauer coincidence experiment that found an asymmetry in a magnetic hyperfine splitting spectrum that is due to parity nonconservation. In Section 7.2. we shall discuss measurements of the multipole-mixing ratio, $E2/M1=\delta$, as well as the phase angle between the $E2$ and $M1$ components. The value of the phase angle has relevance for tests of time reversal.

7.1. Mössbauer Observation of Parity Nonconservation

In order to test the prediction of the nonconservation of parity, Wu et al. [30] in 1957 measured the angular distribution of β particles that were emitted from polarized ^{58}Co and ^{60}Co sources. They found that the intensities, parallel and antiparallel to the axis of polarization, were quite different according to the equation

$$N(\theta) = A(1 + B \frac{v}{c} \cos\theta) \tag{35}$$

where θ is the angle between the gamma ray and the magnetic field H. These authors polarized the radioactive nulei by using very low temperatures and very large magnetic fields ($\mu H \gg kT$).

In the complementary Mössbauer experiment, the nuclei are partially polarized by performing a β–γ coincidence measurement. Since the Wu–Ambler experiments [30] showed that polarized nuclei will yield an asymmetric beta-ray distribution, one would expect that the measurement of beta particles from an unpolarized source would preferentially select nuclear polarizations along the beta axis. The beta measurement would partially polarize the source and give unequal populations to the excited state m^* sublevels

$$P(m^*) = \frac{1}{2I^* + 1}(1 + B' \frac{v}{c} \frac{m^*}{I}) \tag{36}$$

If a large magnetic field is applied in the source along the beta axis, then the hyperfine $m^* \rightarrow m$ transitions will be separated in energy and will be

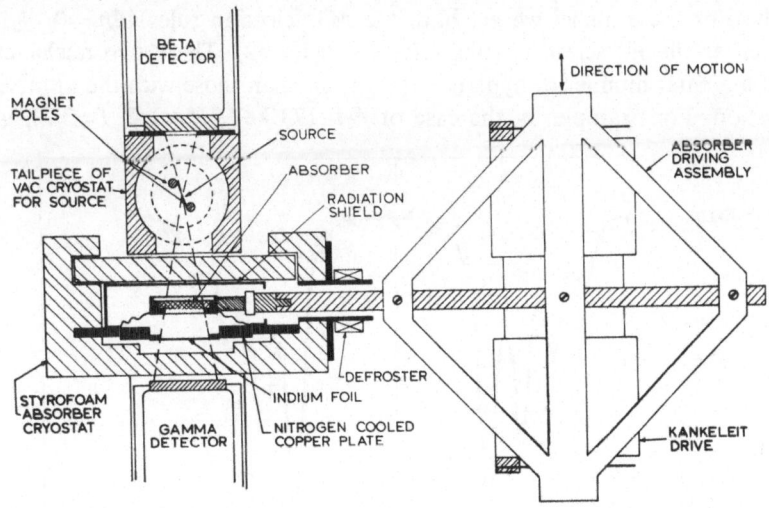

Figure 16. Apparatus used for the β–γ Mössbauer parity experiment from deWaard et al. [24].

measurable by a Mössbauer experiment. By requiring a β–γ coincidence (Figure 16) only those events which have the nucleus partially polarized along the β–γ direction will be considered for Mössbauer analysis.

In a tour de force, this experiment was successfully carried out by deWaard, et al. [24] who used the β^- emitter ^{129m}Te$\rightarrow^{129}$I ($E_{max}=1.556$ MeV, 33 days). An isotope separator was used to substitutionally implant ^{129m}Te into an iron foil, creating a large magnetic field of 1100 kOe at the ^{129}I nucleus. Figure 13 is the spectrum obtained with such a Fe(^{129m}Te) source in a conventional "singles" Mössbauer experiment. For the parity experiment the velocity regions near the two center peaks were counted in coincidence with the beta particles. The two peaks correspond to the parallel and antiparallel orientations of the magnetic moment with respect to the external magnetic field. An asymmetry of 45% was deduced in this experiment and this is in agreement with expectation.

Mössbauer experiments [31, 32] have also been used to set a lower limit on parity violation in strong interactions.

7.2. Multipole Mixing and Time Reversal

Most Mössbauer nuclei emit reasonably pure multipole radiation which is usually either $M1$ for most odd A or odd–odd nuclei, or $E2$ for even–even rotational nuclei. There are, however, at least two cases, ^{99}Ru and ^{193}Ir, where the $E1$ and $M2$ processes compete; $\delta_n{}^2=0.310$ and 2.7, respectively. For these nuclei we get both the $M1$ selection rules ($\Delta m=0, \pm 1$), as well as the $E2$ selection rules ($\Delta m=0, \pm 1, \pm 2$). These two nuclei exhibit a greater number of hyperfine transitions than those with the unmixed radiation. For example, in the case of ^{193}Ir (73 keV, $I^*=1/2$, $I=3/2$), we

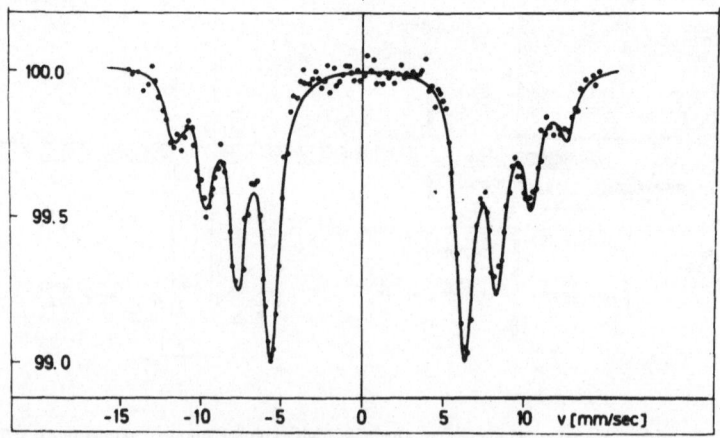

Figure 17. Eight-line spectra of the $M1$-$E2$ mixed 73-keV transition in an Ir–Fe alloy from Wagner et al. [34].

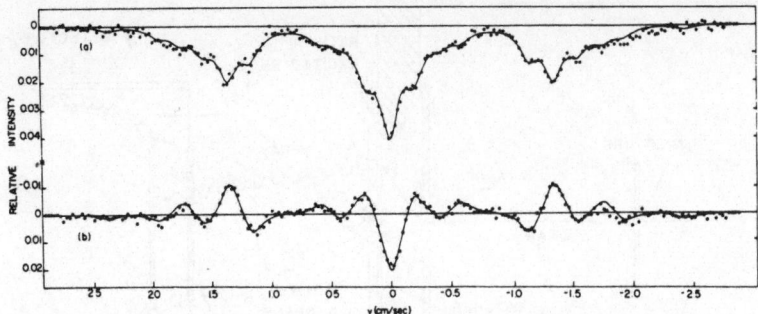

Figure 18. The sum and difference of two ^{193}Ir spectra taken with different magnetic geometries. Such spectra have been used by the Illinois group [35] in the interpretation of the time reversal measurements.

would expect (as in the case of ^{57}Fe) to observe six $M1$ transitions from a cubic magnetic site; but since the 73-keV transition is an $M1$-$E2$ mixed transition there are eight Mössbauer lines (see Figure 17). By using the large magnetic fields of dilute Ir–Fe and Ru–Fe alloys, it has been possible to separate the magnetic hyperfine structure and to measure [33–35] the sign and the magnitude of the mixing coefficient

$$\delta_n = <f\|E2\|i> / <f\|M1\|i> = |\delta|e^{i\eta} \tag{37}$$

to an accuracy of 0.1 %. In these experiments, external magnetic fields have been used to Zeeman-enhance the various Δm transitions.

By proper choice of source and absorber geometries and magnetic fields it is possible to observe $M1$-$E2$ interference terms ($<f|M1|i>$ $<f|E2|i>$) which allows one to set limits on the imaginary part of δ_n. If the phase angle η between the $M1$ and $E2$ matrix elements is not 0 or π, then this would imply that the electromagnetic interaction was not invariant under time reversal. Thus far this small imaginary component has not been observed. Figure 18 is an example of some ^{193}Ir data that has been compiled in the search for this small imaginary component of δ_n. This information comprises the sum and difference of two spectra that were taken with two different magnetic geometries.

8. NUCLEAR REACTIONS AND DEVICES

Since the Mössbauer effect is at the crossroads of nuclear and atomic physics, it is often quite useful for Mössbauer spectroscopists to keep abreast of nuclear technology. Often a particular experimental problem cannot be solved with the conventional radioactive techniques. For example, there are some Mössbauer gamma-ray transitions that are not produced

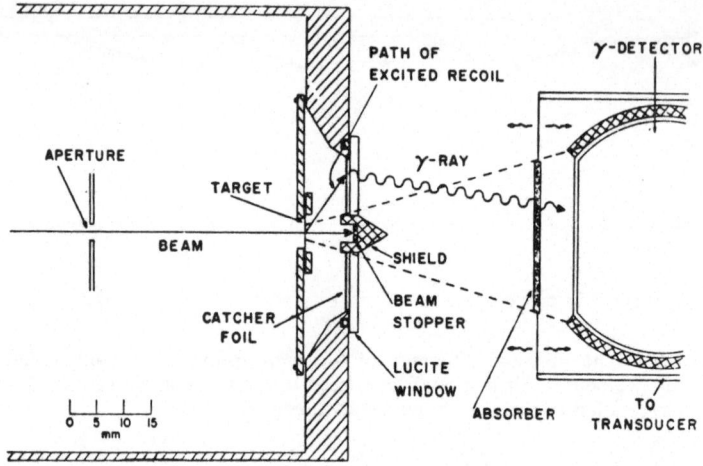

Figure 19. Experimental arrangement that has been used by the Stanford group [36] for Mössbauer studies following coulomb excitation–recoil implantation.

by radioactive decay. For those cases a charged-particle reaction, coulomb excitation, or neutron capture may populate the proper state. The choice of nuclear reaction will depend on cross sections, branching ratios, availability of the appropriate isotopes and radiation damage. Figure 19 shows the coulomb excitation–implantation setup which is used by Hanna

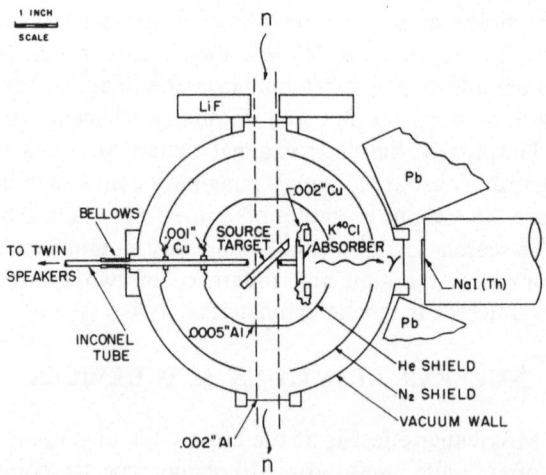

Figure 20. Experimental arrangement for Mössbauer studies following neutron capture from Hafemeister and Shera [37].

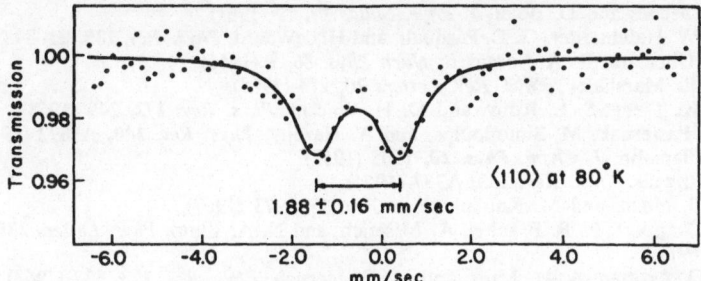

Figure 21. Velocity spectra of ^{57}Co which was implanted in a diamond single crystal with the Argonne isotope separator from Barros *et al.* [38].

and co-workers at Stanford [36]. In this experiment the isotope is coulomb-excited in the target and then implanted by its recoil motion into a substrate.

Heating can be a problem for targets which are made of insulating materials. Thermal neutrons [16, 37] can then be used (Figure 20) to diminish such effects.

Several attempts have been made to measure the Mössbauer effect of ^{57}Co in diamond. The high Debye temperature of diamond along with its simple structure makes it attractive to study. However, diffusion techniques as well as the use of high temperatures and pressures have not produced a suitable source. By using marginal amounts of radioactive isotopes with an isotope separator it is possible to implant the ^{57}Co directly into the diamond and obtain spectra [38] such as that shown in Figure 21.

ACKNOWLEDGMENTS

We are extremely grateful to F. Barros, A. Buffa, A. Klotz, W. Oosterhuis, and J. Vicarro for a critical reading of the manuscript.

REFERENCES

1. D. A. Shirley, *Rev. Mod. Phys.* **36**, 339 (1964).
2. D. W. Hafemeister, *J. Chem. Phys.* **46**, 1929 (1967).
3. A. Gal, L. Grodzins, and J. Hufner, *Phys. Rev. Letters* **21**, 453 (1968).
4. P. Kienle, G. M. Kalvius, and S. L. Ruby, *Hyperfine Structure and Nuclear Reactions*, E. Matthias and D. Shirley, Eds. (North-Holland, Amsterdam, 1968), p. 971.
5. G. K. Shenoy and S. L. Ruby, *Phys. Rev.* **5**, 77 (1970).
6. L. R. Walker, G. K. Wertheim, and V. Jaccarino, *Phys. Rev. Letters* **6**, 98 (1961).
7. R. Ingalls, *Phys. Rev.* **155**, 157 (1967).
8. P. O. Löwdin, *Adv. Phys.* **5**, 1 (1956).
9. W. H. Flygare and D. W. Hafemeister, *J. Chem. Phys.* **43**, 789 (1965).
10. E. Simanek and A. Y. C. Wong, *Phys. Rev.* **166**, 348 (1968).
11. N. Bloembergen and P. Sorokin, *Phys. Rev.* **110**, 865 (1958).

12. M. Menes and D. Bolef, *J. Phys. Solids* **19**, 79 (1961).
13. D. W. Hafemeister, G. DePasquali, and H. deWaard, *Phys. Rev.* **135**, B1089 (1964).
14. R. Uher and R. A. Sorensen, *Nucl. Phys.* **86**, 1 (1966).
15. E. R. Marshalek, *Phys. Rev. Letters* **20**, 214 (1968).
16. P. K. Tseng, S. L. Ruby, and D. H. Vincent, *Phys. Rev.* **172**, 249 (1968).
17. M. Pasternak, M. Simopoulos, and A. Hazony, *Phys. Rev.* **140**, A1892 (1965).
18. R. Bersohn, *J. Chem. Phys.* **20**, 1505 (1952).
19. R. Ingalls, *Phys. Rev.* **133**, A787 (1964).
20. A. J. Nozik and M. Kaplan, *Phys. Rev.* **159**, 273 (1967).
21. J. Chappert, R. B. Frankel, A. Misetich, and N. A. Blum, *Phys. Letters* **28B**, 406 (1969).
22. J. O. Artman, A. H. Muir, and H. Wiedersich, *Phys. Rev.* **173**, 337 (1968).
23. H. deWaard and J. Heberle, *Phys. Rev.* **136**, B1615 (1964).
24. H. deWaard, J. Heberle, P. J. Schurer, H. Hasper, and W. W. J. Koks *Hyperfine Structure and Nuclear Reactions*, E. Matthias and D. Shirley, Eds. (North-Holland, Amsterdam, 1968), p. 329.
25. D. A. Shirley and G. A. Westenberger, *Phys. Rev.* **138**, A170 (1965).
26. G. J. Perlow, W. Henning, D. Olsen, and G. L. Goodman, *Phys. Rev. Letters* **23**, 680 (1969).
27. D. W. Hafemeister and E. B. Shera, *Nucl. Instr. Methods* **41**, 133 (1966).
28. P. Steiner, E. Gerdau, W. Hautsch, and D. Steeken, *Hyperfine Structure and Nuclear Reactions*, E. Matthias and D. Shirley, Eds. (North-Holland, Amsterdam, 1968), p. 364.
29. S. S. Hanna and R. S. Preston, *Phys. Rev.* **139**, A722 (1965).
30. C. S. Wu, E. Ambler, R. W. Hayward, D. D. Hoppes, and R. P. Hudson, *Phys. Rev.* **105**, 1413 (1957).
31. L. Grodzins and F. Genovese, *Phys. Rev.* **121**, 228 (1961).
32. E. Kankeleit, *Proceedings of the International Congress on Nuclear Physics*, Paris, July 1964, Vol. 2, p. 1206.
33. O. C. Kistner, *Hyperfine Structure and Nuclear Reactions*, E. Matthias and D. Shirley, Eds. (North-Holland, Amsterdam, 1968), p. 295.
34. F. Wagner, G. Kaindl, P. Kinele, and H. J. Korner, *Hyperfine Structure and Nuclear Reactions*, E. Matthias and D. Shirley, Eds. (North-Holland, Amsterdam, 1968), p. 187.
35. M. Atac, B. Crisman, P. Debrunner, and H. Frauenfelder, *Phys. Rev. Letters*, **20**, 691 (1968).
36. G. M. Kalvius, G. D. Sprouse, and S. S. Hanna, *Hyperfine Structure and Nuclear Reactions*, E. Matthias and D. Shirley, Eds. (North-Holland, Amsterdam, 1968), p. 686.
37. D. W. Hafemeister and E. B. Shera, *Phys. Rev. Letters* **14**, 593 (1965).
38. F. Barros, D. Hafemeister, and J. Vicarro, *J. Chem. Phys.* **52**, 2865 (1970).

Chapter 4

The Electric Field Gradient Tensor

John C. Travis

National Bureau of Standards
Washington, D.C.

The nuclear portion of the electric quadrupole interaction, along with its application to nuclear physics, was described by Professor Hafemeister in the previous chapter. The importance of the nuclear part, for the purposes of this chapter, is that the extra-nuclear portion, the electric field gradient (EFG) tensor, cannot be extracted from experimental data without prior knowledge of certain nuclear spins and moments. The required information, if known, may be easily located in the *Mössbauer Effect Data Index* [1], and the use of such information to relate observed splittings to the EFG tensor is illustrated in this chapter. In addition, the following sections describe the prediction of the EFG tensor for an assumed molecular crystal model, special techniques and hints, and the utility of EFG information.

The underlying philosophy of the chapter is biased more in the direction of continuity and understandability than theoretical sophistication and completeness. However, a degree of completeness has been introduced by the inclusion of brief references to some exotic considerations with referral to appropriate literature.

1. THE "STANDARD FORM" EFG TENSOR DUE TO A SINGLE POINT CHARGE

The source of the EFG tensor at the nuclear site is the charge-bearing environment of the nucleus. The two principal classifications of this external charge are (1) the electrons directly associated with the nucleus in the atom or ion, and (2) the charges on other ions in the lattice. The simplest and most widely used computational approach, and that used in this chapter, is that of crystal field theory. In this approach, charges external to the central ion are treated as point charges which furnish the so-called "ligand contribution" to the EFG tensor. These charges also perturb the normal free-ion wave function of the central ion, and the electrons occupying the resulting wave function furnish the "valence contribution" to the tensor.

The techniques and conventions applicable to both contributions may be most easily demonstrated by applying them to the simplest example, the ligand contribution of a single point charge. The potential at the Mössbauer nucleus (located at the origin) due to a point charge q at (x, y, z) a distance $r = (x^2 + y^2 + z^2)^{1/2}$ from the origin is given by

$$V = q/r \tag{1}$$

The negative gradient of this potential $-\vec{\nabla} V$, is the electric field $\vec{E}$ at the nucleus, with components

$$E_x = -\frac{\partial V}{\partial x} = qxr^{-3}, \quad E_y = -\frac{\partial V}{\partial y} = qyr^{-3}, \quad E_z = -\frac{\partial V}{\partial z} = qzr^{-3} \tag{2}$$

Finally, the gradient of the electric field at the nucleus $\vec{\nabla}\vec{E}$ is given by

$$\vec{\nabla}\vec{E} = -\vec{\nabla}\vec{\nabla} V = (EFG) = -\begin{bmatrix} V_{xx} & V_{xy} & V_{xy} \\ V_{zx} & V_{yy} & V_{yz} \\ V_{zx} & V_{zy} & V_{zz} \end{bmatrix}, \quad V_{zz} = \frac{\partial^2 V}{\partial x \partial z}, \quad \text{etc.,} \tag{3}$$

where the second partial derivative tensor elements are

$$\begin{aligned}
V_{xx} &= q(3x^2 - r^2)r^{-5} \equiv U_{11}, \\
V_{yy} &= q(3y^2 - r^2)r^{-5} \equiv U_{22}, \\
V_{zz} &= q(3z^2 - r^2)r^{-5} \equiv U_{33}, \\
V_{xy} &= V_{yx} = 3qxyr^{-5} \equiv U_{12}, \\
V_{xz} &= V_{zx} = 3qxzr^{-5} \equiv U_{13}, \\
V_{yz} &= V_{zy} = 3qyzr^{-5} \equiv U_{23}
\end{aligned} \tag{4}$$

The U symbols defined in Eq. (4) make possible a more compact notation convenient for computer programming:

$$\begin{aligned}
U_{ij} &= q(3x_i x_j - r^2\delta_{ij}), \quad x_1 = x, \ x_2 = y, \ x_3 = z \\
\delta_{ij} &= 0 \text{ if } i \neq j, \ \delta_{ij} = 1 \text{ if } i = j
\end{aligned} \tag{5}$$

The EFG elements may be expressed in spherical polar coordinates as

$$r = (x^2 + y^2 + z^2)^{1/2}, \quad \theta = \cos^{-1}(z/r), \quad \phi = \tan^{-1}(y/x) \tag{6}$$

by making the substitutions

$$x = r\sin\theta\cos\phi, \quad y = r\sin\theta\sin\phi, \quad z = r\cos\theta \tag{7}$$

into Eq. (4) to give

$$
\begin{aligned}
V_{xx} &= q(3\sin^2\theta\cos^2\phi - 1)r^{-3} \\
V_{yy} &= q(3\sin^2\theta\sin^2\phi - 1)r^{-3} \\
V_{zz} &= q(3\cos^2\theta - 1)r^{-3} \\
V_{xy} &= V_{yx} = 3qr^{-3}\sin^2\theta\sin\phi\cos\phi \\
V_{xz} &= V_{zx} = 3qr^{-3}\sin\theta\cos\theta\cos\phi \\
V_{yz} &= V_{zy} = 3qr^{-3}\sin\theta\cos\theta\sin\phi
\end{aligned}
\tag{8}
$$

Of the nine components of the EFG tensor, only five may be considered to be independent parameters. Three of the off-diagonal elements (U_{ij}, $i \neq j$) are dependent since $V_{xy} = V_{yx}$, etc., and one of the diagonal elements (U_{ij}, $i=j$) is dependent because Laplace's equation

$$\nabla^2 V = V_{xx} + V_{yy} + V_{zz} = 0 \tag{9}$$

must be satisfied.

The values of the EFG tensor elements obviously depend upon the choice of the coordinate axes. For this reason, a "standard form" has been designated for the EFG tensor, defining a unique set of axes known as "the principal axes of the EFG tensor." This unique coordinate system is the one for which the off-diagonal elements are zero and the diagonal elements are ordered, such that

$$|V_{xx}| \leq |V_{yy}| \leq |V_{zz}| \tag{10}$$

It is often necessary to mathematically manipulate a trial EFG tensor based on an arbitrary axis set in order to identify the principal axes. The initial axes will be distinguished from the principal axes in this chapter by the association of italics with the former (x, y, z, U) and roman (upright) type (x, y, z, U) with the latter. Thus, Eq. (10) refers to the principal axes (cf. Appendix I).

With the tensor in standard form, only two of the five independent parameters of the EFG tensor seem to be left. In reality, however, the three independent off-diagonal elements have been replaced by the three Euler angles (α, β, γ) necessary to describe the relative orientation of the principal and initial axes. The orientation parameters are often neglected since they rarely (only for simultaneous internal magnetic and quadrupole interactions) affect line positions in Mössbauer spectra, and only for mixed interactions or single crystal spectra do they affect the line intensities. They should be measured wherever possible, though, to give more parameters for correlation with proposed theoretical models.

By convention, the two independent parameters used to describe the diagonal elements are $V_{zz}(U_{33})$ and the asymmetry parameter, η, defined by

$$\eta \equiv \frac{V_{xx} - V_{yy}}{V_{zz}} = \frac{U_{11} - U_{22}}{U_{33}} \tag{11}$$

Equation (10) imposes the restriction that

$$0 \leq \eta \leq 1 \tag{12}$$

The mathematical procedure alluded to earlier for determining the principal axes is the process of matrix diagonalization. A relatively simple procedure for diagonalizing third-order matrices (such as the EFG tensor) is described in Appendix I. However, for the present single-point-charge illustration, a few educated guesses will be sufficient to diagonalize the matrix and find the standard form parameters. To begin with, notice that if any one of the three initial axes passes through the charge, the off-diagonal elements are automatically zero because of their proportionality to products such as xy. Thus, if we choose to pass the x axis through our charge, giving it coordinates $(r, 0, 0)$, then the diagonal EFG elements, by Eq. (4) are given by

$$V_{xx} = 2qr^{-3}, \qquad V_{yy} = qr^{-3}, \qquad V_{zz} = -qr^{-3} \tag{13}$$

In order to satisfy Eq. (10), take the old x axis to be the new z axis, yielding

$$V_{xx} = -qr^{-3}, \qquad V_{yy} = -qr^{-3}, \qquad V_{zz} = 2qr^{-3} \tag{14}$$

or

$$V_{zz} = 2qr^{-3} \qquad \eta = 0 \tag{15}$$

The x and y axes were not uniquely defined by Eqs. (10) and (14) for this example. Indeed, these axes are only uniquely defined for cases for which $\eta \neq 0$, due to the fact that $\eta = 0$ reflects at least three-fold rotational symmetry about the z axis.

Equation (15) is consistent with the point-charge example in Chapter 3 for $q = -e$.

2. THE LIGAND CONTRIBUTION

The portion of the EFG tensor which results from all of the other charged ions in the lattice is sometimes called the "lattice contribution," but is probably more widely known as the "ligand contribution." The significance of the latter designation is in the r^{-3}-dependence of the EFG tensor, which gives much more importance to ions directly coordinated to the central ion, i.e., to the ligands, than to more distant ions in the lattice.

On the other hand, complete neglect of distant ions may lead to inaccuracies by the sheer force of numbers, since the number of ions in a volume shell of thickness dr at a distance r is proportional to r^2.

The EFG tensor for a collection of charges is the element-wise sum of the individual EFG tensors. Thus, for a collection of n charges, Eq. (5) becomes

$$U_{ij} = \sum_{k}^{n} q_k (3 x_{ki} x_{kj} - r_k^2 \delta_{ij}) \tag{16}$$

where q_k is the charge and (x_{k1}, x_{k2}, x_{k3}) is the position of the kth ion. As for the single charge, point-charge calculations of the ligand contribution are particularly simple if rectangular Cartesian coordinates can be oriented in such a way that each ion lies on an axis. Just as before, this procedure yields a diagonal matrix, making only the axis permutation to satisfy Eq. (10) necessary to put the tensor into standard form.

As an example, consider the *cis* and *trans* isomers of the hypothetical octahedrally coordinated compound MA_2B_4, where M denotes the

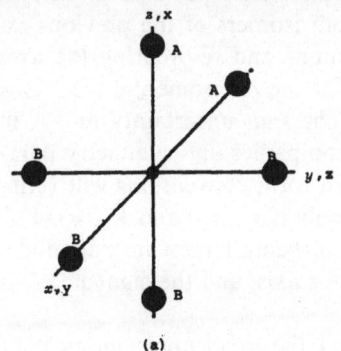

(a)

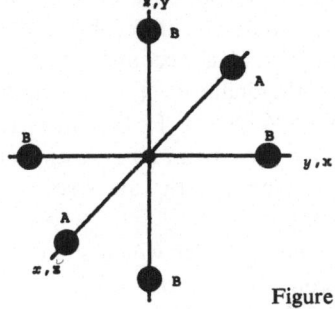

(b)

Figure 1. (a) *Cis* and (b) *trans* isomers of the hypothetical system MA_2B_4.

Table 1.   Ligand Contributions for *Cis*- and *Trans*-MA$_2$B$_4$

Ligand	V_{xx}	*Cis* V_{yy}	V_{zz}	V_{xx}	*Trans* V_{yy}	V_{zz}	Ligand
B(+x)	$2S_b$	$-S_b$	$-S_b$	$2S_a$	$-S_a$	$-S_a$	A(+x)
A(−x)	$2S_a$	$-S_a$	$-S_a$	$2S_a$	$-S_a$	$-S_a$	A(−x)
B(+y)	$-S_b$	$2S_b$	$-S_b$	$-S_b$	$2S_b$	$-S_b$	B(+y)
B(−y)	$-S_b$	$2S_b$	$-S_b$	$-S_b$	$2S_b$	$-S_b$	B(−y)
A(+z)	$-S_a$	$-S_a$	$2S_a$	$-S_b$	$-S_b$	$2S_b$	B(+z)
B(−z)	$-S_b$	$-S_b$	$2S_b$	$-S_b$	$-S_b$	$2S_b$	B(−z)
Total	S_a-S_b	$2S_b-2S_a$	S_a-S_b	$4S_a-4S_b$	$2S_b-2S_a$	$2S_b-2S_a$	

Mössbauer atom. Figure 1 illustrates these isomers and shows in italics the arbitrarily chosen initial axes, used for the calculations in Table 1. The table also employs the definitions $S_a \equiv q_A r_A^{-3}$, and $S_b \equiv q_B r_B^{-3}$.

Equation (10) may be satisfied by relabelling the axes as shown by the roman type axis labels in Figure 1, with the result that $V_{zz} = 2S_b - 2S_a$ and $\eta = 0$ for the *cis* isomer, while $V_{zz} = 4S_a - 4S_b$ and $\eta = 0$ for the *trans* isomer.

As another interesting example, the *cis* and *trans* isomers of MA$_3$B$_3$ may be examined by replacing B(+y) in both isomers of the previous example with A(+y). Summing the table columns and re-ordering the axes then yields the result that $V_{zz} = 0$ and $\eta = 0$ for the *cis* isomer, while $V_{zz} = \pm 3(S_b - S_a)$ and $\eta = 1$ for the *trans* isomer. The sign uncertainty in V_{zz} in the *trans* case is a feature which always accompanies an asymmetry parameter of unity. Consideration of the standard form conventions will verify that the asymmetry parameter can be one only if $V_{xx} = 0$ and $V_{yy} = -V_{zz}$. Since two of the elements tie for the honor of being largest in magnitude, either of the two axes may be taken to be the z axis, and the sign of V_{zz} is indeterminate.

The interaction between the ligands and the quadrupole moment of the Mössbauer nucleus is complicated by the presence of the electronic cloud belonging to the Mössbauer atom. Sternheimer [2] has shown that, for relatively large atoms, this cloud distorts in the presence of a ligand quadrupole interaction in such a way as to amplify the interaction. The effect is thus called "Sternheimer antishielding" and is accounted for by multiplying the elements of the EFG tensor by the quantity $(1 - \gamma_\infty)$, where γ_∞ is the "Sternheimer antishielding factor." With the EFG tensor in standard form, only V_{zz} is thus modified, since the asymmetry parameter is a ratio. The antishielding factor may be quite large, with values such as $\gamma_\infty = -10.6$ for Fe^{2+} [3] and -9.14 for Fe^{3+} [4].

As will be shown later, there are circumstances under which the total EFG tensor is comprised of only the ligand contribution. In such "ligand-only" cases, the standard form EFG parameters calculated as in this section should be reflected in the observed quadrupole coupling. For the more general case, however, it would be premature to put the tensor into standard form before including the valence contribution.

3. THE VALENCE ELECTRON CONTRIBUTION

In order to evaluate the contribution of the valence electrons to the EFG tensor, it is first necessary to extend the point charge formalism to continuous charge distributions. Equation (16), which gives the EFG elements for a collection of point charges, may be expressed for a continuous distribution as

$$U_{ij} = \int_\tau \rho(x,y,z) \; (3x_i x_j - \delta_{ij}) r^{-5} \; d\tau \tag{17}$$

where the point charges have been supplanted by charge elements $\rho d\tau$, with ρ being the charge density and $d\tau$ the volume element, and the sum has been extended into an integral over the volume of the distribution. Similar expressions, such as

$$V_{zz} = \int_\tau \rho(r,\theta,\phi) \; (3\cos^2\theta - 1) r^{-3} \; d\tau \tag{18}$$

may be written for the spherical polar forms, with the volume element $d\tau$ being given by $r^2 \sin\theta dr d\theta d\phi$, instead of $dxdydz$.

A case of particular interest is that of a spherically symmetric charge distribution, for which $\rho(r, \theta, \phi)$ becomes $\rho(r)$ and Eq. (18) yields

$$V_{zz} = \int_0^\infty \rho(r) \; r^2 \; dr \int_0^{2\pi} d\phi \int_0^\pi (3\cos^2\theta - 1) \; \sin\theta d\theta - 0 \tag{19}$$

from the θ integral alone. Similar results may be obtained for all of the other EFG elements. Thus, the spherically symmetric portion of the atomic charge cloud does not contribute to the EFG tensor. For this reason, only the electrons above a full shell configuration (the valence electrons) need be considered as *direct* contributors to the EFG tensor. All of the electrons contribute *indirectly* to the EFG tensor by means of Sternheimer shielding and antishielding.

The EFG elements for the valence electrons may be obtained by replacing the density in Eq. (18) with its quantum mechanical equivalent, the square of the wave function. For instance, consider a system with one electron beyond a spherical core, with this electron occupying a hydrogenic d_{xy} orbital given by

$$\psi_{xy} = (15/2)^{1/2} \; \sin^2\theta \sin 2\phi \tag{20}$$

Table 2. EFG Elements for the d Orbitals

	d_{xy}	d_{xz}	d_{yz}	$d_{x^2-y^2}$	d_{z^2}
$V_{xx}/(q<r^{-3}>)$	2/7	2/7	−4/7	2/7	−2/7
$V_{yy}/(q<r^{-3}>)$	2/7	−4/7	2/7	2/7	−2/7
$V_{zz}/(q<r^{-3}>)$	−4/7	2/7	2/7	−4/7	4/7

The element V_{xx} for this case is given by

$$V_{xx} = (15/2)q <r^{-3}> \int_0^{2\pi} \int_0^{\pi} (\sin^2\theta\sin2\phi)^2(3\sin^2\theta\cos^2\phi - 1)$$
$$\times \sin\theta d\theta d\phi = -(2/7)e<r^{-3}>, \tag{21}$$

where $q = -e$ is the charge on the electron, and the value of the radial integral is denoted symbolically as the expectation value of r^{-3}. Similar integrals may be evaluated for the remaining EFG elements for this orbital, as well as for the other orbitals. Table 2 gives the results of the angular integrations for the "stationary"[1] d orbitals. The off-diagonal elements all are zero.

Notice that the EFG elements are already ordered according to Eq. (10) for our electron in a d_{xy} orbital. Had the electron instead been in a d_{xz} orbital, however, an axis rotation would have been required. With the appropriate axis relabelling, four of the five d orbitals ($d_{x^2-y^2}, d_{xy}, d_{xz}, d_{yz}$), taken independently, yield identical EFG tensors. This is quite reasonable inasmuch as these four orbitals have exactly the same shape and differ only in orientation. Thus, for our example, or for one electron in any of the four identical d orbitals, the standard form EFG parameters are $V_{zz} = (4/7)e<r^{-3}>$ and $\eta=0$, where $q=-e$ is the charge on the electron. The results for a single vacancy, or "hole," in an otherwise full d manifold, differ only in sign from the single electron results, because the charge to be considered is $q=+e$.

The foregoing example of an electron in a d_{xy} orbital could rigorously occur only at a temperature of absolute zero. At other temperatures, each of the d orbitals would be partially populated, due to thermal energy, in accordance with the Boltzmann equation

$$P_i = \exp(-E_i/kT) \left[\sum_j^n \exp(-E_j/kT)\right]^{-1} \tag{22}$$

where T is the absolute temperature, k is the Boltzmann constant of 0.694 cm^{-1} deg^{-1} or 1.38×10^{-16} erg deg^{-1}, E_i is the energy of the ith of a total of

---

[1] As distinguished from the "rotating" solutions to the free hydrogen atom problem, d_0, d_{+1}, and d_{+2}. The stationary forms are generally a more convenient starting point for bound atom calculations.

n levels, and P_i is the fractional population of the ith level by the electron. From Eq. (22) it may be seen that $\sum\limits_{j}^{n} P_j = 1$.

The ratio of the population of the ith and the jth levels may be seen from Eq. (22) to be

$$P_i/P_j = \exp(-E_i/kT)/\exp(-E_j/kT)$$
$$= \exp(-\Delta E/kT), \ \Delta E = E_i - E_j \qquad (23)$$

From this equation it is apparent that the energy reference point, or zero, is unimportant, as long as all energies are referred to the same point. It is normally convenient to refer to the lowest energy level, or ground state, as zero energy. In this reference frame, the ratio of the population of the ith energy level to that of the ground state is given by

$$P_i/P_0 = \exp(-E_i/kT) \qquad (24)$$

For a system at temperature T, the values of the EFG elements are given by the weighted averages of the contributions $(U_{ij})_k$ due to each of the states. The weighting factor to be used is the relative population P_k. Thus,

$$U_{ij} = \sum_{m=1}^{n} P_m(U_{ij})_m$$
$$= \sum_{m=1}^{n} (U_{ij})_m \ \exp(-E_m/kT)[\sum_{m=1}^{n} \exp(-E_m/kT)]^{-1} \qquad (25)$$

The upper limit on the number of energy levels to be used, n, is set arbitrarily. Strictly speaking, every energy level has a nonzero population. If, however, we arbitrarily decide to limit the summation to states which have at least one hundredth the population of the ground state, we impose the inequality

$$0.01 < \exp(-E_i/kT) \qquad (26)$$

or

$$E_i < \ln(100)kT = 4.606kT \qquad (27)$$

Thus, we would confine the sum to all of the energy levels lying below 4.606 kT, referred to the ground state.

As an example, consider a d^1 system with nominally tetrahedral symmetry, but with a slight distortion causing the $d_{x^2-y^2}$ orbital to lie an energy Δ_t above the d_{z^2} orbital, as shown in Figure 2a. Assuming the t_{2g} levels to lie about 4000 cm^{-1} above the e_g orbitals in energy, we may use the criterion of Eq. (27) to show that the t_{2g} orbitals contribute a negligible amount to the tensor for temperatures below

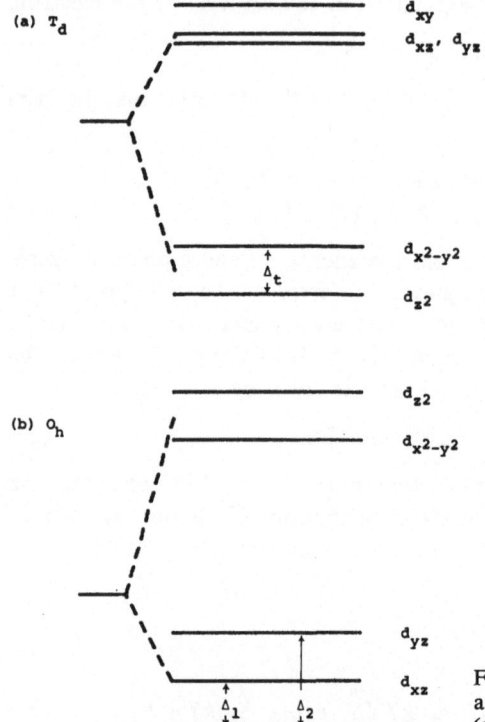

Figure 2. Electronic configuration for a single d electron in slightly distorted (a) tetrahedral symmetry and (b) octahedral symmetry.

$$T = 4000 \text{ cm}^{-1}/4.606 \times 0.694 \text{ cm}^{-1} \text{ deg}^{-1} = 1252°\text{K} \qquad (28)$$

Using Table 1 for the partial EFG element contributions of the d orbitals, and populating only the e_g orbitals with one electron, gives

$$V_{xx} = [(-e<r^{-3}>) (-2/7)\exp(0) + (-e<r^{-3}>) (2/7)\exp(-\Delta_t/kT)]$$
$$\times [\exp(0) + \exp(-\Delta_t/kT)]^{-1}$$
$$V_{yy} = V_{xx}, \qquad (29)$$
$$V_{zz} = (+4/7) (-e<r^{-3}>) [\exp(0) - \exp(-\Delta_t/kT)]$$
$$\times [\exp(0) + \exp(-\Delta_t/kT)]^{-1}$$

with the d_{z^2} orbital as the zero of energy. The formulation would still be correct for $d_{x^2-y^2}$ lying low, but Δ_t would then have to be a negative number. By taking the zero of energy halfway between the two e_g orbitals, Eq. (29) may be reduced to hyperbolic tangent functions.

In this example, the absolute magnitudes of the valence EFG elements depend on temperature, but the relative magnitudes are fixed. In cases where orbitals with different symmetry axes are involved, the relative magnitudes vary as well. For instance, for a d^1 system in orthorhombically distorted octahedral symmetry, as illustrated in Figure 2b, the EFG elements would be given by

$$V_{xx} = (2/7) \; (-e\!<\!r^{-3}\!>) \; [1 + \exp(-\Delta_1/kT) - 2 \; \exp(-\Delta_2/kT)]/B$$
$$V_{yy} = (2/7) \; (-e\!<\!r^{-3}\!>) \; [1 - 2\exp(-\Delta_1/kT) + \exp(-\Delta_2/kT)]/B \quad (30)$$
$$V_{zz} = (-2/7) \; (-e\!<\!r^{-3}\!>) \; [2 - \exp(-\Delta_1/kT) - \exp(-\Delta_2/kT)]/B$$
$$B \equiv 1 + \exp(-\Delta_2/kT) + \exp(-\Delta_2/kT)$$

where only the t_{2g} orbitals have been considered.

In addition to cases with only one electron or one hole in the valence shell, one-electron theory may be applied to cases for which only one electron or hole is free to move as a function of temperature. As examples, consider the three iron configurations illustrated in Figure 3. In both of the high-spin d^6 (ferrous) cases, none of the five ($\uparrow$) electrons may be

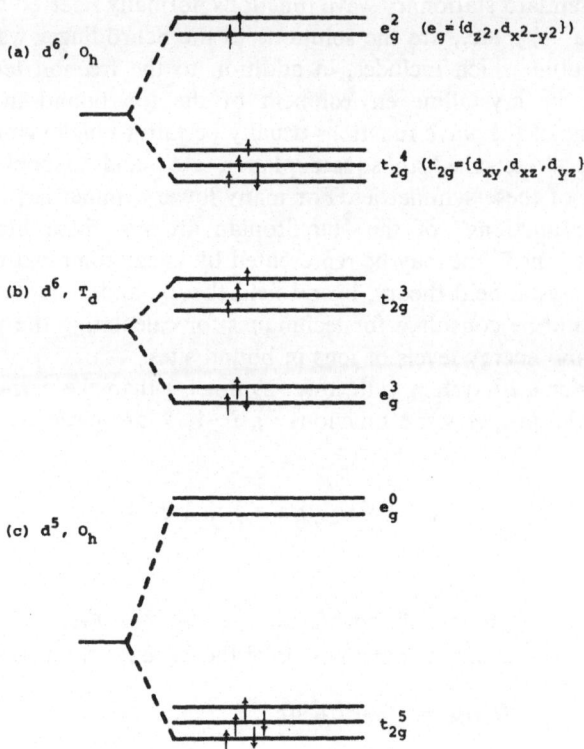

Figure 3. Three iron configurations suitable for one-electron method: (a) high-spin octahedral Fe^{2+}, (b) high-spin tetrahedral Fe^{2+}, and (c) low-spin octahedral Fe(III).

moved without changing the net spin, and therefore only the one ($\downarrow$) electron may move with thermal energies. The contribution of the five ($\uparrow$) electrons to the EFG tensor is zero since they, as a group, have the same symmetry (spherical) as the full d shell. This can also be verified by summing the contributions of one electron in each of the orbitals, i.e., summing the rows of Table 2. In the low-spin d^5 (ferric) case shown, only the hole in the otherwise full t_{2g} level is free to move without changing the net spin of the system. The configuration is equivalent to three ($\uparrow$) electrons, three ($\downarrow$) electrons, and one ($\downarrow$) hole. By adding the contributions to the EFG elements of two electrons in each of the t_{2g} orbitals, using Table 2, it may be seen that the six electrons together contribute nothing to the EFG tensor. Thus, only the hole, of charge $q = +e$, is responsible for the tensor for this configuration. The one-electron results in Eqs. (29) and (30) may thus be applied directly to the two ferrous examples and, with the change of only the sign, to the ferric example.

The standard stationary wave functions normally referred to in chemistry, such as d_{xy}, etc., are the solutions to the Schrödinger wave equation for a potential which includes, in addition to the free-ion terms, a term describing the crystalline environment of the ion bound in the lattice. These forms of the wave functions usually pertain to high symmetries such as octahedral, tetrahedral, square planar, etc., and to some symmetric distortions of these symmetries. For many lower symmetries, however, the true "eigenfunctions" of the Hamiltonian are not these high-symmetry "basis functions," but may be represented by linear combinations of them. Books on crystal field theory, ligand field theory, and/or molecular orbital theory should be consulted for techniques for calculating the proper wave functions and energy levels of ions in bound sites.

Consider a d^1 system with lower symmetry than the earlier examples, such that the proper wave functions ψ_i, $i = 1, 5$, are given by linear combinations

$$\psi_i = \sum_{j}^{5} a_{ij} d_j \qquad i, j = 1, 5 \tag{31}$$

where

$$\{d_j, j = 1, 5\} = \{d_{x^2-y^2}, d_{xz}, d_{z^2}, d_{yz}, d_{xy}\} \tag{32}$$

The EFG tensor element contributions of the wave function ψ_k are given by

$$(U_{ij})_k = \int_{\tau} |\psi_k|^2 U_{ij} \, d\tau$$

$$= \sum_{m=1}^{5} \sum_{n=1}^{5} a_{km}^* a_{kn} \int_{\tau} d_m^* U_{ij} d_n \, d\tau \tag{33}$$

The remainder of the problem of predicting the temperature-dependent

Table 3. Values of the Integrals $<d_k|U_{ij}|d_k,>$ in Units of $2q<r^{-3}>/7$ [a]

$<d_k|$	V_{xx}	V_{yy}	V_{zz}	V_{xy}	V_{xz}	V_{yz}	$|d_{k'}>$
$<d_{x^2-y^2}|$	0	0	0	0	1	0	$|d_{xz}>$
$<d_{x^2-y^2}|$	$-\sqrt{3}$	$\sqrt{3}$	0	0	0	0	$|d_{z^2}>$
$<d_{x^2-y^2}|$	0	0	0	0	0	-1	$|d_{yz}>$
$<d_{xz}|$	0	0	0	0	$\sqrt{3}/2$	0	$|d_{z^2}>$
$<d_{xz}|$	0	0	0	3/2	0	0	$|d_{yz}>$
$<d_{xz}|$	0	0	0	0	0	3/2	$|d_{xy}>$
$<d_{z^2}|$	0	0	0	0	0	$\sqrt{3}/2$	$|d_{yz}>$
$<d_{z^2}|$	0	0	0	$-\sqrt{3}$	0	0	$|d_{xy}>$
$<d_{yz}|$	0	0	0	0		3/2	$|d_{xy}>$

[a] The values for $k=k'$ are given in Table 2.
Integrals not shown in either table are zero.

EFG tensor, given that the energies corresponding to the ψ_i are known, proceeds as before.

Once the integrals $\int_\tau d_m{}^*U_{ij}d_n \, d\tau$ in Eq. (33) have been tabulated, only the coefficients a_{km} are required to calculate the EFG elements for a d^1 (or high-spin d^6) system. Notice that the values of the integrals for $m=n$ are just those given in Table 2. In addition, the cross product terms, as reported by Bielefeld [5], are recorded in Table 3.

The techniques described above may be applied to systems for which the one-electron approach is inadequate by using the free-ion wave functions as basis functions, e.g., the ^3F functions for d^8. Instead of Boltzmann-populating orbitals with an electron, one populates the states of the system with the total charge. Tables 2 and 3 may be applied to systems with D states since these have the same angular dependence as hydrogenic d orbitals.

As for the ligand contribution, the valence contribution to the EFG tensor at the nucleus is not that due to the valence electrons alone, but includes also the distortion of the normally spherical core electrons. The distortion differs from that caused by the ligands inasmuch as the valence electrons actually overlap the core electrons. This distortion is of such a sense as to slightly diminish the effective EFG tensor seen by the nucleus and is thus referred to as Sternheimer shielding. The effective EFG elements are obtained by multiplying the tensor elements by $1-R$, where R is the valence electron Sternheimer shielding factor.

4. QUADRUPOLE SPLITTINGS

It is evident from the preceding discussions that the total effective EFG tensor seen by the Mössbauer nucleus is given by

$$U_{tj} = (1 - \gamma_\infty)\ (U_{tj})_{ltg} + (1 - R)\ (U_{tj})_{val} \qquad (34)$$

It is implicitly assumed in Eq. (34) that all three EFG tensors involved are expressed with respect to the same set of real space axes. One must avoid the temptation to diagonalize the component tensors individually before adding them, inasmuch as this generally leads to the meaningless operation of adding tensors having different real space bases.

Once the total effective EFG tensor Eq. (34), has been determined for a given model, and the parameters V_{zz} and η have been determined by matrix diagonalization, the remaining step is to relate these parameters to observable spectral parameters through consideration of the nuclear quadrupole interaction for the nuclide of interest.

4.1. The Nuclear Quadrupole Interaction

The method for finding the quadrupolar perturbations of the nuclear energy levels will be discussed in Dirac's nomenclature, utilizing the operator relationships

$$<Im'|Im> = \delta_{mm}, \qquad \hat{I}_z|Im> = m|Im>$$
$$\hat{I}^2|Im> = I(I+1)|Im> \qquad \hat{I}_\pm|Im> = \sqrt{(I\mp m)\ (I\pm m+1)}|I\ m\pm1>$$
$$\qquad (35)$$

Since two nuclear energy levels are involved, the method must generally be applied both to the excited state, of spin I_e, and to the ground state, of spin I_g.

The initial step for a level of spin I is to find all $(2I+1)^2$ matrix elements of the form

$$<Im|\hat{H}|Im'> \qquad m,m' = -I,\ -I + 1,\ \ldots\ldots,\ I - 1,\ I \qquad (36)$$

where H is the nuclear quadrupole Hamiltonian operator

$$\hat{H} = \frac{eQV_{zz}}{4I(2I - 1)}[3\hat{I}_z^2 - \hat{I}^2 + \frac{\eta}{2}(\hat{I}_+^2+\hat{I}_-^2)] \qquad (37)$$

The elements are then arranged in a square matrix (array) with rows and columns labelled by the corresponding values of m and m'. The arrangement of the values of m and m' is arbitrary, but the two must be ordered identically. As an example, Table 4 shows the Hamiltonian matrix for the excited state, $I_e=3/2$ for ^{57}Fe, with the arbitrary ordering $\{3/2,\ -1/2,\ -3/2,\ 1/2\}$ for m and m'.

If the matrix is diagonal, the diagonal elements are the quadrupole perturbation energies of the corresponding m substates. For instance, the Hamiltonian matrix of Table 4 is diagonal for cases for which $\eta=0$, and,

Table 4. Excited State ($I=3/2$) Hamiltonian Matrix for ^{57}Fe[a]

$m' \rightarrow$ $m \downarrow$	3/2	$-1/2$	$-3/2$	1/2
3/2	3	$\sqrt{3}\eta$	0	0
$-1/2$	$\sqrt{3}\eta$	-3	0	0
$-3/2$	0	0	3	$\sqrt{3}\eta$
1/2	0	0	$\sqrt{3}\eta$	-3

[a] In units of $eQV_{zz}/12$.

for such cases, the $m=\frac{3}{2}$ and $m=-\frac{3}{2}$ substates are both perturbed in energy by an amount $E(\pm\frac{3}{2})=3eQV_{zz}/12$, and the $m=\frac{1}{2}$ and $m=-\frac{1}{2}$ substates are both perturbed by $E(\pm\frac{1}{2})=-3eQV_{zz}/12$. If the matrix is not initially diagonal, the perturbation energies may be found by the process of matrix diagonalization. Diagonalization of matrices larger than 3×3 is usually done by iterative methods on digital computers, but some may be "blocked" into matrices small enough to be diagonalized analytically. The example of Table 4 may be blocked as indicated by the dotted lines, and the two 2×2 matrices diagonalized independently. The results for this example are $E'(\pm\frac{3}{2})=E(\pm\frac{3}{2})\ (\sqrt{1+\eta^2/3})$ and $E'(\pm\frac{1}{2})=E(\pm\frac{1}{2})$ $(\sqrt{1+\eta^2/3})$, where the labels used were chosen to be appropriate in the limit of $\eta\rightarrow0$.

The number of lines in the experimental quadrupole split spectrum for a given Mössbauer nuclide is determined by the number of energy levels of each of the two states and by the selection rules for the gamma radiation. Continuing with the ^{57}Fe example, the ground state ($I_g=\frac{1}{2}$) has no quadrupole moment ($eQ = 0$) and therefore is unsplit. The radiation is known to be magnetic dipole, which has the selection rules $\Delta m=0, \pm1$. Thus, transitions are allowed between the ground state and both levels of the excited state, as illustrated in Figure 4, yielding the familiar ^{57}Fe two-line quadrupole split spectrum. The observed splitting of the two lines is given by

$$\Delta E_Q = |E'(\pm\tfrac{3}{2}) - E'(\pm\tfrac{1}{2})| = |eQV_{zz}(\sqrt{1 + \eta^2/3})/2| \qquad (38)$$

The single quadrupole splitting parameter for ^{57}Fe, ΔE_Q, is defined as a positive quantity because the two lines are indistinguishable for normal powder samples, and hence the measurement of the sign associated with V_{zz} requires special techniques. Another obvious difficulty associated with quadrupole studies of ^{57}Fe systems is the impossibility of extracting both V_{zz} and η from the single quadrupole splitting. Techniques for overcoming these difficulties will be discussed in Section 5. For the purposes of this section, however, it is important to note that a number of Mössbauer nuclides, having at least one state of $I>\frac{3}{2}$, have at least two quadrupole

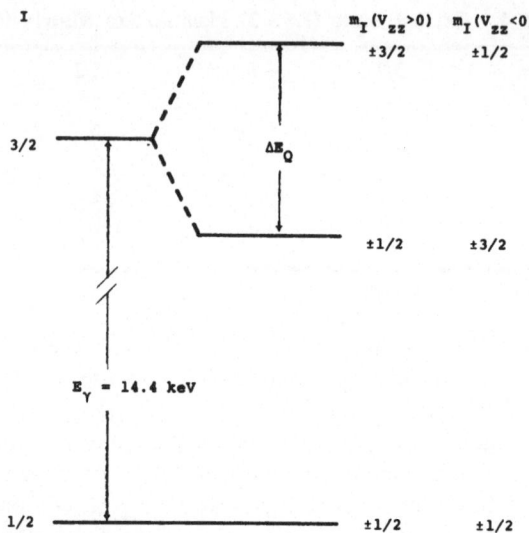

Figure 4.   Nuclear energy level scheme for quadru-
pole-split ^{57}Fe, for $\eta = 0$.

splittings (i.e., at least three lines) which are different functions of V_{zz} and η, such that these parameters may be measured directly from the normal spectra. Most of the rare-earth Mössbauer nuclides, for instance, fall into this category.

The nuclear theory outlined above may now be coupled with the EFG calculations of the previous section to calculate the expected quadrupole splitting behavior for particular cases. For convenience of discussion, it is desirable to categorize cases as "ligand-only," "valence-only", or "ligand–valence combined." In the limit of pure spherical symmetry in the valence electrons, e.g., high-spin Fe^{3+}, Eq. (34) reduces to the "ligand-only" case.

4.2.   Ligand-Only Splittings

As noted in the ligand contribution section, the EFG elements for ligand-only cases contain linear combinations of the "ligand strength" parameters $S_i = q_i < r_i^{-3} >$ with each element amplified by the factor $1 - \gamma_\infty$. The matrix diagonalization process may be legitimately simplified by factoring out all of the common multiplicative factors first and then reintroducing them into the resulting eigenvalues. Thus, the Sternheimer antishielding term may always be factored, and sometimes terms which are functions of the ligand strengths may be factored from the tensor in ligand-only cases.

For ligand strength studies, it is convenient to subdivide the ligand-only case into three distinct possibilities: (1) when only V_{zz} depends on the relative ligand strengths—the asymmetry parameter and the principal axes, including their name designations, are independent of the ligand strengths; (2) V_{zz} and η depend on the relative ligand strengths, but the principal axes, excluding the name designations, do not; and (3) all parameters are functions of the ligand strengths.

Possibility number one is well illustrated by the examples given earlier for the ligand contribution to the EFG. From Table 1 it may be seen that the term $(S_a - S_b)$ can be factored from both of the cis- and trans-MA_2B_4 tensors. The indicator for this possibility is that all of the ligand strength dependence can be factored, leaving only pure numbers in the tensor. Thus, even if the tensor is not diagonal in the initial axes, the diagonalization is unique, not affected by the ligand strengths, and the principal axes are also unique. The uniqueness of the asymmetry parameter follows from the fact that it is a ratio of linear combinations of the diagonal elements, and, therefore, any term factorable from the tensor cancels in the ratio.

If the Mössbauer nuclide M is Fe^{3+}, the cis- and trans-MA_2B_4 quadrupole splittings could be calculated from Table 1 and Eq. (38), yielding

$$\Delta E_Q(cis) = |eQ(1 - \gamma_\infty)\ (S_b - S_a)|$$
$$\Delta E_Q(trans) = |2eQ(1 - \gamma_\infty)\ (S_a - S_b)| \tag{39}$$

Thus, the normally observed quadrupole splitting differs by a factor of two for the two geometrical isomers. Detection of the change in sign of V_{zz} would require special techniques described later.

The second possibility may be illustrated with a hypothetical square planar compound trans-MA_2B_2. By substituting zeros for the $B(+z)$ and $B(-z)$ contributions to the EFG tensor in Table 1, the results

$$V_{xx} = 4S_a - 2S_b, \qquad V_{yy} = 4S_b - 2S_a, \qquad V_{zz} = -2S_a - 2S_b \tag{40}$$

may be obtained. Note that since the initial axes were chosen so as to pass through the ligands the tensor remains diagonal regardless of the ligand strengths. However, the relative strengths do determine the labelling of the axes to satisfy the ordering convention, Eq. (10). Assuming S_a and S_b to have the same sign (probably negative), then

$$
\begin{array}{llll}
V_{xx} = V_{yy}, & V_{yy} = V_{zz}, & V_{zz} = V_{xx}, & \text{for } |S_b| \ll |S_a| \\
V_{xx} = V_{xx}, & V_{yy} = V_{zz}, & V_{zz} = V_{yy}, & \text{for } |S_a| \ll |S_b| \\
V_{xx} = V_{yy}, & V_{yy} = V_{xx}, & V_{zz} = V_{zz}, & \text{for } |S_b| \gtrsim |S_a| \\
V_{xx} = V_{xx}, & V_{yy} = V_{yy}, & V_{zz} = V_{zz}, & \text{for } |S_a| \gtrsim |S_b|
\end{array} \tag{41}
$$

Thus, for the last case, $|S_a| \gtrsim |S_b|$, the EFG parameters are

$$V_{zz} = -2S_a - 2S_b \qquad \eta = 3(S_b - S_a)/(S_b + S_a) \tag{42}$$

and the splitting, for the case of ^{57}Fe, would be

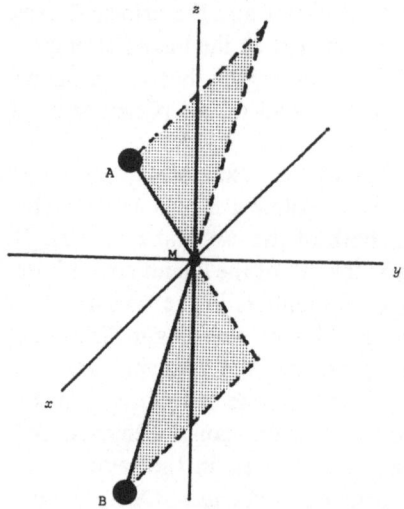

Figure 5. Hypothetical compound MAB. The z axis of the EFG falls somewhere in the shaded surface, depending on the relative ligand strengths.

$$\Delta E_Q = eQ(1-\gamma_\infty)|S_a + S_b|\{1 + 9(S_b - S_a)^2/(S_a + S_b)^2\}^{1/2} \quad (43)$$

A simple example of the third possibility is the unlikely hypothetical compound MAB, where the bond angle AMB is 120°, as illustrated in Figure 5. Choosing the axes shown in the figure as initial axes yields

$$\theta_A = 30°, \quad \phi_A = 0, \quad \theta_B = 150°, \quad \phi_B = 0$$
$$V_{xx} = 5(S_a + S_b)/4, \quad V_{yy} = -(S_a + S_b), \quad V_{zz} = -(S_a + S_b)/4 \quad (44)$$
$$V_{xz} = 3\sqrt{3}(S_a - S_b)/4$$

with the remaining EFG elements being zero. Diagonalization of the EFG tensor would show the variability of all of the parameters with varying ligand strengths, but this characteristic can be more simply illustrated by considering three special cases. The limiting and points $|S_a| \gg |S_b|$ and $|S_b| \gg |S_a|$ may be approximated by the single charge EFG model of Eq. (15) by letting $S_b \to 0$ and $S_a \to 0$, respectively. Thus, for the former, the z axis passes through ligand A, $V_{zz} = 2S_a$, and $\eta = 0$, while for the latter, the z axis passes through B, $V_{zz} = 2S_b$, and $\eta = 0$. For the intermediate case, $S_a = S_b$, $V_{xz} = 0$, from Eq. (44), and thus $V_{zz} = (5/4)S_a$, $\eta = 3/5 = 0.6$, with the principal axes being the initial axes.

For any other choice of ligand strengths, diagonalization would be required, yielding a principal z axis in the shaded region of Figure 5, for S_a and S_b of the same sign. Although Eq. (44) may be diagonalized analytically, the resulting expression for the quadrupole splitting as a function of the relative ligand strengths would be too obscure to be very useful. For this reason, numerical treatment, with graphs of V_{zz}, η, and quadrupole splitting(s) as functions of the strengths resulting, is recommended for compounds of this category. For the present example, with iron or tin, graphs of V_{zz} and ΔE_Q, in units of $eQ(1-\gamma_\infty)S_a$, and of η, as functions of S_b/S_a, would be useful.

It is obvious from the point-charge EFG equations that ligand-only quadrupole splitting studies can be concerned not only with ligand strengths (i.e., effective values of q/r^3) but with bond angles as well. Bond-angle studies, just as type three ligand strength studies, generally require diagonalization, numerical treatment, and graphical presentation of the results as a function of bond angle(s).

Theoretical ligand-only quadrupole splitting studies are generally presented on a relative scale instead of an absolute one, due, in part, to the difficulty in assigning effective q/r^3 values (particularly to covalently bonded ligands), and, in part, to the uncertainty in the reported values of eQ and of γ_∞ for the various nuclides. Indeed, ligand-only studies may be effective for nuclides for which eQ and γ_∞ have not been reported. It should be noted that the asymmetry parameter is unitless, and when it is experimentally measurable, offers a direct, absolute comparison between experimental and theoretical results.

4.3. Valence-Only Splittings

As noted earlier, the valence contribution to the EFG tensor arises when the valence electrons aspherically populate the valence shell. Such a departure from spherical symmetry generally occurs only when the ligands are distorted from pure cubic symmetry. Thus, strictly speaking, truly "valence-only" splittings are a highly unlikely occurrence. However, the valence-only *approximation* is widely used in the interpretation of experimental results, inasmuch as the valence charge is closer to the nucleus than the ligands, and the valence contribution is thus, in general, greater.

Valence splitting is only an indirect probe of molecular structure, but it is a powerful tool for examining the quantum mechanical configuration of the central ion. The necessary model for valence-splitting calculations is not so much a structural model as a quantum mechanical model, with the splitting being expressed as a function of the valence orbital energies, wave function coefficients, spin-orbit coupling constant, etc., in addition to the temperature. Of course, the central ion configuration is in turn related to the molecular structure through the potential function employed in the Hamiltonian.

As an example of a valence-only splitting calculation, consider a slightly distorted tetrahedral Fe^{2+} system such as is treated in Figure 3b and Eq. (29). By inspection it may be seen that the EFG elements given in Eq. (29) satisfy the ordering requirement [Eq. (10)] such that the initial axes are also the principal axes. Since the asymmetry parameter is zero regardless of temperature, the value of V_{zz} from Eq. (29) substituted into the iron splitting expression [Eq. (38)] yields as the quadrupole splitting

$$\Delta E_Q = |e^2 Q(1 - R)\ (2 < r^{-3} > /7)| \frac{1 - \exp(-\Delta_t/kT)}{1 + \exp(-\Delta_t/kT)} \tag{45}$$

where the Sternheimer factor $(1-R)$ has been included in the final expression for V_{zz}. As mentioned earlier, Eq. (45) may be expressed as a hyperbolic tangent function by shifting the zero of energy by $\Delta_t/2$. Experimental data on several compounds of this type at several temperatures have been reported by Edwards et al. [6].

The splitting expression for an orthorhombically distorted octahedral Fe^{2+} system, for which the EFG elements are given in Eq. (30), would be considerably more complicated, since the asymmetry parameter would be nonzero for temperatures above absolute zero. For such systems, it would be particularly advantageous to be able to monitor the asymmetry parameter experimentally.

In the previous section on ligand-only splittings, it was mentioned that theoretical quadrupole splitting expectations are usually presented on a relative scale. This implies that such studies must involve more than one compound, or an asymmetry parameter measurement, to be meaningful. In other words, any calculated result could be perfectly matched with any experimental result by an appropriate choice of scaling factor, unless the asymmetry parameter is measured. Valence-only quadrupole splitting calculations are also generally expressed on a relative scale. In addition to containing the inaccurately known quadrupole moment(s) and Sternheimer shielding factor R, the valence splitting "scaling factor" contains the expectation value of r^{-3} over the valence wave function. This factor may differ significantly from reported "free-ion" calculations for various reasons, including covalency and the nephelauxetic effect. Thus, Eq. (45) might be more usefully expressed as

$$\Delta E_Q(T) = \Delta E_Q(0) \frac{1 - \exp(-\Delta t/kT)}{1 + \exp(-\Delta_t/kT)} \tag{46}$$

Such an expression could be least-squares fitted to temperature-dependent experimental results with $\Delta E_Q(0)$ and Δ_t as fitting parameters (see, for instance, Ref. 6).

By comparing fitted zero-temperature limit values for different compounds of a given Mössbauer nuclide, one may, in effect, measure the relative values of $<r^{-3}>$ for the various compounds. The other fitted parameter(s), the valence orbital splitting(s), may be related to a structural model through a quantum mechanical treatment, and may be compared with corresponding splittings of companion compounds, or with values determined by other experimental methods, such as electron paramagnetic resonance (EPR). Thus, valence splitting calculations may be meaningfully related to a single compound but may be even more useful when related to a family of compounds.

4.4. Ligand-Valence Combined Splittings

The rationale behind the common "valence-only" approximation becomes apparent when one considers the complexity of the complete combined interaction problem. To begin with, factors such as $(1-\gamma_\infty)qr^{-3}$ and $(1 - R)<r^{-3}>$, which could be conveniently factored before diagonalizing and reintroduced as empirically evaluated scaling factors in the ligand- and valence-only cases, cannot be factored from the complete tensor of Eq. (34). Furthermore, the principal axes of the ligand and valence contributions to the EFG are not necessarily collinear, so that the EFG parameters and principal axis directions may be rather transdendental functions of the relative magnitudes of the contributions.

There are some special cases for which the problem reduces from a tensor to a scalar one. Such is the case, for instance, for systems in which both contributions are individually diagonal with respect to a single set of coordinate axes. In such a case, the total tensor is also diagonal with respect to this basis, regardless of the relative ligand and valance contributions. One classic treatment of this sort of problem in the literature is given by Ingalls [7]. For his purposes, he was able to express the final effective EFG parameters as

$$V_{zz} = (1 - \gamma_\infty) \ (V_{zz})_{lig} + (1 - R) \ (V_{zz})_{val}$$
$$\eta V_{zz} = (1 - \gamma_\infty)\eta_{lig}(V_{zz})_{lig} + (1 - R)\eta_{val}(V_{zz})_{val} \tag{47}$$

In general, however, one would need to examine the relative magnitudes of the resultant EFG parameters before assigning axis designations.

The complexity of the theoretical description of the behavior of the quadrupole splitting for the combined interaction may be a blessing or a curse, depending on the interest of the researcher. The combined interaction problem requires more parameters than either component interaction, meaning, of course, that more experimental data are needed, particularly on the temperature dependence curve. On the brighter side, however, they also give more flexibility in the fitting of data, perhaps explaining "anomalous" results obtained with valence-only fitting of the data.

To try to encapsulate a completely generalized "cookbook" procedure for interpreting mixed interaction problems would be a serious mistake, in addition to being a difficult undertaking. One could conceivably incorporate all of the parameters necessary to describe the ligand contribution (for the assumed structural model), plus all of the parameters necessary to describe the valence contribution (for the assumed configuration of valence electrons), plus three additional Euler angle parameters (to describe the relative orientation of the two tensors), plus a matrix diagonalization routine, plus the equation(s) for calculating the splitting(s) from the EFG elements, all into a nonlinear least-squares curve-fitting computer program for application to temperature-dependent data. Such a procedure may indeed fit data, but it loses track of the physical relationship between the two contributions and therefore loses some important constraints. The relationship in question is just this: from the point of view of crystal field theory, the potential due to the ligands, whose various second partial derivatives constitute the EFG tensor, is the same potential which determines the quantum mechanical configuration of the valence electrons. Inclusion of this relationship into combined interaction studies, when possible, results in a reduction of the number of parameters. Unfortunately, the description of this relationship in a usably parameterized form may itself be a formidable task. Again, the work of Ingalls [7], utilizing such a ligand–valence relationship for tetragonally distorted octahedral Fe^{2+} systems, is a good example.

A foreseeable development in the state of the art of interpreting quadrupole splittings would be the incorporation into curve-fitting programs of theoretical quadrupole splitting prediction programs that require only estimated ligand positions and strengths as input parameters. In other words, the necessary quantum mechanics, Boltzmann populating, and so forth, would be done internally. At least one theoretical splitting program, appropriate for such an application, has been written for Fe^{2+} systems by Bielefeld [5]. Even this approach, however, has its admitted drawbacks. Not the least of these is the fact that crystal field theory is at present the only feasible quantum mechanical method to use, making the application to covalent systems risky at best. Most of the difficulties are of a developmental nature, such as the needed refinement in values of quadrupole moments, Sternheimer factors, expectation values of r^{-3}, etc., and the improvement of existing computer techniques.

The above remarks on the combined EFG interaction should illuminate the current tendency toward utilization of the ligand- or valence-only approximations wherever possible, and the relegation of the combined interaction problem to qualitative discussions.

5. POTPOURRI

This section is included to deal with a handful of loose ends which have limited application but are convenient to be aware of when appropriate cases arise. Particular attention is given to ^{57}Fe because of the popularity of this nuclide and the difficulty of obtaining complete EFG information for it.

5.1. Compounds with Internal Magnetic Fields

The extraction of quadrupole splitting parameters from an experimental spectrum is often complicated by the superposition of a magnetic hyperfine interaction. The appearance of the spectrum for such a "lavesphase" compound is governed not only by the magnitudes of the EFG parameters and the internal magnetic field, but also by the orientation of the magnetic field with respect to the EFG principal axes. Such data are often interpreted by visual comparison with curves generated by theoretical computer programs such as that of Gabriel and Ruby [8]. In fact, such theoretical routines have been incorporated into curve-fitting programs in several laboratories. If one of the interactions is appreciably larger than the other, first-order perturbation theory may be conveniently applied, as described by Wertheim [9]. The first-order result for ^{57}Fe, for a small quadrupole perturbation on a large magnetic splitting, is that the quantity

$$(\Delta E_Q/2)\ (3\ \cos^2\alpha - 1)V_{zz}/|V_{zz}| \tag{48}$$

may be extracted from experimental data. Here, α is the angle between the magnetic field and the principal axis of the EFG tensor, and ΔE_Q is defined as a positive quantity. Thus, the elusive sign of V_{zz} is almost revealed in laves-phase iron spectra but is, in fact, still masked by the presence of the angular factor.

5.2. Applied Magnetic Fields

As previously mentioned, one of the primary irritations of studying quadrupole effects in ^{57}Fe is the presence of only one measurable parameter in powdered, zero-magnetic field samples. Ruby and Flinn [10] first suggested the possibility of applying an external field in order to display more information in the spectrum. Theoretical studies (analytical, to first order by Collins [11], and exact, numerical by Gabriel and Ruby [8]) verified that the two normally identical lines become markedly different in the presence of the field. The fact of being able to identify the lines means that the sign of V_{zz} may be measured in this way. In addition, the fine structure of the perturbed spectra may be examined to yield rough values of the asymmetry parameter, within limits determined by magnetic anisotropy in paramagnetic compounds and by the vibrational isotropy of the sample [12].

In the limit of zero asymmetry parameter and for applied fields of about 25 kOe or more, the ^{57}Fe $m_I=\pm\frac{1}{2}$ line assumes the appearance of a triplet and the $m_I=\pm\frac{3}{2}$ line appears as a doublet. A convenient mnemonic for determination of the sign of V_{zz} for ^{57}Fe is that the doublet occurs at the more positive velocity for V_{zz} positive, and vice versa. As the asymmetry parameter increases, the lines begin to look more alike, to the point of becoming identical triplets for an asymmetry parameter of unity. Although the exact appearance of the lines depends on whether the field is parallel or perpendicular to the experimental axis (the direction of the incident radiation), the rough behavior described above applies to either case.

5.3. The Gol'danskii-Karyagin Effect

There are two known reasons for the occasional observation of an intensity asymmetry in the two lines of a ^{57}Fe quadrupole split doublet in a powdered sample. Both effects are dependent upon the fact that the intensities for a single crystal absorber are, in general, not equal. The most obvious cause of intensity asymmetry in a powdered sample, then, is simply preferential orientation, and the cure is improved sample grinding, or mixing the sample with some noninterfering powder, such as chalk dust. If the asymmetry cannot be "cured," it may result from the second, more theoretically sophisticated and also more useful, effect.

Karyagin [13] has derived an expression for the ratio of the two lines as a function of the difference of the mean-square vibrational amplitudes of the nucleus along and perpendicular to the V_{zz} axis. Since the recoilless fraction is related to the mean-square vibrational amplitude in the direction of the incident gamma, the contribution of each microcrystal in the sample is dependent upon its orientation, yielding a "preferential intensity," having an effect similar to preferential orientation.

The Gol'danskii–Karyagin effect may, on occasion, be employed to deduce the sign of V_{zz} in both ^{57}Fe and ^{119m}Sn spectra. Specifically, the documented [14] dependence of the intensity ratio upon $(<z^2> - <x^2>)$ may be employed to identify the lines, thus determining the sign of V_{zz}, if the sign of $(<z^2> - <x^2>)$ is known. Conversely, the sign of the vibrational anisotropy factor can be determined if the sign of V_{zz} is known. In iron and tin, the $m_I = \pm 3/2$ line is the more intense line for negative values of the anisotropy parameter $(<z^2> - <x^2>)$ and is the less intense for positive values.

5.4. Single Crystal Samples

From the preceding discussion, it is apparent that one may add an additional measurable parameter, namely the peak intensity ratio, to the quadrupole split spectra of iron and tin compounds by the use of single crystals. However, to obtain accurate values of the EFG parameters, one must have several differently oriented crystal slices, with the orientation carefully and precisely measured, say, by x-ray methods.

Zory [15] has expressed the relative intensities of the two lines as

$$I_{\pm}(\theta,\phi) = N_{\pm}^{2}[\lambda_{\pm}^{2}(1 + \cos^{2}\theta)/2 + (5 - 3\cos^{2}\theta)/6$$
$$+ (\lambda_{\pm}/\sqrt{3})\cos2\phi(1 - \cos^{2}\theta)] \qquad (49)$$
$$\lambda_{\pm} = [\sqrt{3} \pm \sqrt{3 + \eta}]/\eta, \qquad N_{\pm}^{2} = (1 - \lambda_{\pm}^{2})^{-1}$$

where I_{+} is the intensity of the $m_{I}=\pm\frac{3}{2}$ line and I_{-} is the intensity of the $m_{I}=\pm\frac{1}{2}$ line.

6. THE UTILITY OF EFG INFORMATION

Traditionally, quadrupole splitting has been used most often in the literature in a nonquantitative fashion, as a sort of measure of the distortion of a system from spherical symmetry. In concluding this chapter, it is appropriate to consider (1) the pitfalls of the traditional approach, (2) the justifiable rationale for the traditional approach, and (3) the hope for the emergence of meaningful and quatitative systematics for the use of EFG information in the future.

One of the foremost pitfalls of the traditional method is in the understanding of "distortion," as applied to quadrupole splitting data. For instance, for the ligand-only splitting of the high-spin Fe^{3+} cis-$MA_{2}B_{4}$ system, given in Eq. (39), the distortion would be determined by $S_{b}-S_{a}$, since the splitting is proportional to this quantity. Note that the same quadrupole splitting could be shared by two different compounds, one having ($S_{b}=2$ units, $S_{a}=1$ unit, $S_{a}/S_{b}=0.5$) and the other having ($S_{b}=4$ units, $S_{a}=3$ units, $S_{a}/S_{b}=0.75$). Thus, although the "strength" ratios differ markedly, the compounds would have the same "distortion" as measured by Mössbauer spectroscopy.

Another important pitfall of the traditional method is the necessity of recognizing the effects of combined ligand–valence splittings. As an example, consider high-spin Fe^{2+} in slightly tetragonally distorted octahedral symmetry. Ingalls [7] has shown that the ligand and valence contributions to V_{zz} invariably have opposite signs for such a system. Thus, at the low-temperature limit, where the valence contribution depends on neither distortion nor temperature, increased ligand distortion would decrease V_{zz} and hence the quadrupole splitting, since the valence contribution would be dominant. At the high-temperature limit, the problem would be ligand-only and be subject to the same difficulties discussed above, but would, within those restrictions, correspond to the general principle that increased quadrupole splitting indicates increased distortion. In the intermediate temperature range, where the valence contribution is both temperature- and distortion-sensitive, the story becomes very complicated. The splittings of the d orbitals, which determine the valence contribution through their Boltzmann electronic populations, respond to a q/r type of distortion, whereas the ligand contribution to the EFG responds to a q/r^3 type of distortion. Since both contributions increase, in opposite senses, in response to two different types of "distortion," the meaningful assignment of a distortion–quadrupole splitting relationship for such systems at intermediate temperatures requires some prior assumptions.

The above remarks are not intended to belittle those who have used, or will use, the qualitative approach in interpreting quadrupole splitting data. It is hoped, however, that most of the pitfalls will become more widely appreciated and understood in the Mössbauer community. The rationale behind the past and continuing use of qualitative methods may be readily understood from a brief review of this chapter. The most obvious difficulty is that a given set of measured EFG parameters does not correspond uniquely to a single system configuration. For instance, two ions, each of charge q, located at coordinate positions $(0, 0, +a)$ and $(0, 0, -a)$ yield the same EFG parameters as a single ion of charge $2q$ located at either position. Another good example is that given earlier in this section for cis-MA_2B_4. Obviously, then, structural confirmation by measurement of EFG parameters is a tricky business. In general, one goes by the circuitous route of refuting proposed models, hoping that only one will be compatible with the data in the final analysis.

One may also cite, in justification of the qualitative approach, the restrictive computational requirements of quantitative treatment, the difficulties and inadequacies of molecular quantum mechanics, the inaccuracies in reported moments and shielding factors, and so forth. The interesting point about this latter list of problems is that all of them are subject to improvement and, in fact, *are* improving. Good computational facilities are becoming widespread, some of the larger facilities are making possible significant breakthroughs in molecular quantum mechanics (EFG measurements should, in fact, contribute to the refinement of molecular quantum mechanics), and nuclear moment and shielding factor calculations are improving with improved facilities and theory. Another encouraging trend is the growing number of laboratories equipped for temperature studies.

Mössbauer spectroscopy is entering a new era of sophistication, based on the technical and theoretical developments of the last decade. This era is destined to be accompanied by increasingly imaginative and quantitative applications of EFG information.

APPENDIX I

Diagonalization of the EFG Tensor

The EFG tensor may be diagonalized analytically by solving the "secular equation"

$$\begin{vmatrix} V_{xx}-\lambda & V_{xy} & V_{xz} \\ V_{yx} & V_{yy}-\lambda & V_{yz} \\ V_{zx} & V_{zy} & V_{zz}-\lambda \end{vmatrix} = 0 \qquad (I\text{-}1)$$

The expansion of the determinant may be simplified by the use of LaPlace's equation, $V_{xx}+V_{yy}+V_{zz}=0$, and the symmetry of the tensor, yielding the resultant third-degree equation

$$\lambda^3 + a\lambda + b$$
$$a = V_{yy}V_{zz} + V_{xx}V_{zz} + V_{xx}V_{yy} + V_{xy}{}^2 + V_{yz}{}^2 + V_{xz}{}^2 \qquad (I\text{-}2)$$
$$b = V_{xy}{}^2V_{zz} + V_{xz}{}^2V_{yy} + V_{yz}{}^2V_{xx} - V_{xx}V_{yy}V_{zz} - 2V_{xy}V_{yz}V_{xz}$$

If the roots of Eq. (I-2) are not obvious for a particular case, they may be determined by the equations

$$\lambda_i = 2\sqrt{-a/3}\,\cos[\phi/3 + 120(i-1)] \qquad (I\text{-}3)$$
$$i = 1,3 \qquad \cos\phi \equiv -b/(2\sqrt{-a^3/27}$$

The difference between the original EFG tensor and the diagonalized tensor is that they are related to different coordinates; that is, the "basis vectors" have been rotated. To find the orientation of the new coordinates with respect to the old, it is necessary to find the "eigenvectors" of the EFG. This is done by solving three systems of three homogeneous equations in three unknowns

$$(V_{xx} - \lambda_i)X_i + V_{xy}Y_i + V_{xz}Z_i = 0$$
$$V_{yx}X_i + (V_{yy} - \lambda_i)Y_i + V_{yz}Z_i = 0 \qquad (I\text{-}4)$$
$$V_{zx}X_i + V_{zy}Y_i + (V_{zz} - \lambda_i)Z_i = 0$$

where the λ_i have previously been determined by Eq. (I-3). The quantities X_i, Y_i, and Z_i represent the projections on the old coordinate axes of a unit vector along the ith new coordinate axis. Since the equations are homogeneous, they can only be solved for two of the unknowns in terms of the third. The third value is established by requiring that the vector be normalized such that $X^2{}_i + Y^2{}_i + Z^2{}_i = 1$. Suppose for a given problem that λ_1 is established by the absolute magnitude criterion Eq. (10) to be V_{yy}. Then a unit vector $\mathbf{m}$ along the new y axis would be given by

$$\mathbf{m} = X_1\mathbf{i} + Y_1\mathbf{j} + Z_1\mathbf{k} \qquad (I\text{-}5)$$

Table II-1

Configu-ration	Sym-metry	$\Delta E_Q(0)$	Temperature dependence	Comments
Fe^{2+} high-spin	O_h	large	yes	One electron above half-full, use t_{2g} only
	T_d	large	yes	Same, only use e_g only
Fe(II) low-spin	O_h	large	no	Full t_{2g}, empty e_g; ligand part large due to strong bonding, nephelauxetic effect
	T_d	large	yes	No known occurrences; would be two electrons in t_{2g}
Fe^{3+} high-spin	O_h	small	no	Half-full shell, ligand part small
	T_d	small	no	Same
Fe(III) low-spin	O_h	intermediate	yes	Hole in t_{2g}; ligand part unusually large due to strong bonding, may partially cancel valence part
	T_d	intermediate	yes	Same except one electron in t_{2g}; ligand and valence parts may partially cancel; no known occurrences

where $\mathbf{i}$, $\mathbf{j}$, $\mathbf{k}$ and $\mathbf{l}$, $\mathbf{m}$, $\mathbf{n}$ are unit vectors along the old and new coordinate axes, respectively.

APPENDIX II

Expected Splitting Behavior for Representative Iron Configurations

The oversimplified qualitative Table II-1 is intended to relate the material of this chapter to some real systems, and give the reader an opportunity to test his understanding of the chapter by verifying the table. The symmetries are all assumed to be slightly distorted, as they invariably are in real systems.

REFERENCES

1. A. H. Muir, Jr., K. J. Ando, Helen M. Coogan, Eds., *Mössbauer Effect Data Index 1958–1965* (Interscience Publishers, New York, 1966).
2. R. M. Sternheimer, *Phys. Rev.* **80**, 102(1950); **84**, 244(1951); **130**, 1423(1963); H. M. Foléy, R. M. Sternheimer, and D. Tycko, *ibid.* **93**, 734(1954); R. M. Sternheimer and H. M. Foley, *ibid.* **102**, 731(1956).
3. R. Ingalls, *Phys. Rev.* **128**, 1155(1962).
4. R. M. Sternheimer, *Phys. Rev.* **130**, 1423(1963).
5. Michael J. Bielefeld, *Semi-empirical Treatment of the Mössbauer Quadrupole Split-*

ting, Research Institute for Natural Sciences, Woodstock College, Woodstock, Maryland, NASA grant NsG-670.

6. P. R. Edwards, C. E. Johnson, and R. J. P. Williams, *J. Chem. Phys.* **47**, 2074(1967).
7. R. Ingalls, *Phys. Rev.* **133**, A787(1964).
8. . J. R. Gabriel and S. L. Ruby, *Nucl. Instr. Methods* **36**, 23(1965).
9. G. K. Wertheim, *Mössbauer Effect* (Academic Press, New York, 1964), p. 80.
10. S. L. Ruby and P. A. Flinn, *Rev. Mod. Phys.* **36**, 351(1964).
11. R. L. Collins, *J. Chem. Phys.* **42**, 1072(1965).
12. J. C. Travis, Ph.D. thesis, Department of Physics, The University of Texas, Austin, Texas, August, 1967; R. L. Collins and J. C. Travis, *Mössbauer Effect Methodology*, Vol. 3, (Plenum Press, New York, 1967), p. 123.
13. S. V. Karyagin, *Dokl. Akad. Nauk SSSR* **148**, 1102(1963) (in Russian); *Proc. Acad. Sci. USSR, Phys. Chem. Sec.* **148**, 110(1964) (in English).
14. P. A. Flinn, S. L. Ruby, and W. L. Kehl, Scientific Paper 63-128-117, p.6, Westinghouse Research Laboratories, Beulah Road, Churchill Borough, Pittsburgh, Pennsylvania.
15. P. Zory, *Phys. Rev.* **140**, A1401(1965).

Chapter 5

Application to Solid-State Physics

Robert L. Ingalls

University of Washington
Seattle, Washington

This chapter will concentrate on some of the more general and fundamental relationships between the nucleus and the surrounding electrons as well as between the Mössbauer effect atom and its neighbors.[1]

Thus, we are really concerned with the atomic parts of the various electric and magnetic hyperfine interactions, i.e., charge density at the nucleus $\psi^2 (0)$, effective magnetic field at the nucleus H, or electric field gradient (EFG) tensor components at the nucleus $-V_{ij}$. Whether you actually call these terms solid-state properties, rather than chemical or metallurgical, etc., depends upon your point of view. In addition, of course, we shall be concerned with some of the dynamical aspects such as lattice vibrations and relaxation phenomena. After all, the Mössbauer effect, recoilless emission and absorption, is itself a solid-state effect.

For simplicity most of the examples will concern Mössbauer effect studies with ^{57}Fe. This perhaps is justified if you remember that a large fraction of all the Mössbauer effect experiments do in fact deal with ^{57}Fe. Moreover, study of the solid-state aspects of ^{57}Fe experiments extends readily to Mössbauer experiments with other isotopes because iron is found in all types of systems from conducting to insulating, ferromagnetic to diamagnetic, and ionic to covalent.

1. ISOMER SHIFT

Let me remind you again what the isomer shift is. We consider the basic Mössbauer gamma ray from the transition between the nuclear excited state

---

[1] Most of the references for this chapter may be found in the excellent book by G. K. Wertheim, *Mössbauer Effect* (Academic Press, New York, 1964).

and the ground state. Its total energy can be divided up into a pure nuclear part plus another part that depends on the size of the nucleus and the density of s electrons at the nucleus. For the source S it is written

$$\Delta E_S = \frac{2}{5}\pi Ze^2(R_{\text{ex}}^2 - R_{\text{gnd}}^2)\,\psi_S^2(0) \tag{1}$$

and represents a difference in electrostatic binding to the electrons. Equation (1) is the simple nonrelativistic expression for a uniformly charged nucleus; the more accurate relativistic expression should of course be used in actual application. Thus, we have a small part of the gamma-ray energy that really depends upon the environment. Of course, in our Mössbauer experiment we also have ΔE_A for the absorber that we match with ΔE_S of the source. The isomer shift, $\delta = \Delta E_A - \Delta E_S$, is just the energy by which this source gamma ray must be augmented in order to be absorbed. (We neglect temporarily the thermal shift.)
Thus

$$\delta = \frac{2\pi}{5}Ze^2(R_{\text{ex}}^2 - R_{\text{gnd}}^2)\,(\psi_A^2(0) - \psi_S^2(0)) \tag{2}$$

or, writing it in another way,

$$\delta = \alpha\psi_A^2(0) + \text{const} \tag{3}$$

This is a convenient form if we have a standard source, but we look at many

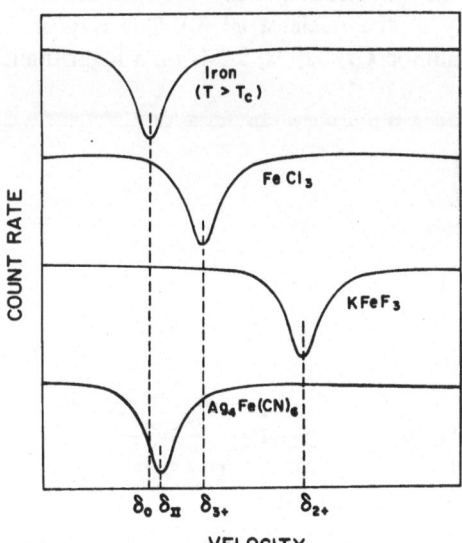

Figure 1. Isomer shifts for different ^{57}Fe absorbers.

absorbers. Then the source will just contribute a constant, and the isomer shift will be a linear function of the charge density ψ^2 (0) in the absorber.

Now let us go into some of the experimental results with ^{57}Fe. A typical source, for instance ^{57}Co in copper, would consist of a nonsplit single line. The absorbers which we examine can be quite varied (Figure 1). One type would just be a metal, for instance metallic iron. I'll call that Fe_0, meaning that it is almost neutral atomic iron in some respects, and neglect all other interactions. (We know iron is ferromagnetic at room temperature but pretend it is above its Curie temperature.) Other classes would include the ionic compounds of iron such as $FeCl_3$, where iron is in the 3+ state. If we were to run a Mössbauer spectrum of this, we would find another single line, but it would be at a higher velocity. Or, there is another ionic state (2+) that iron likes quite well, ferrous, such as in $KFeF_3$. If we were to run a Mössbauer spectrum of such a compound, we would find an even greater isomer shift.

In a highly covalent iron compound like $Ag_4Fe(CN)_6$, yet another isomer shift is observed. In other words, as we vary the charge on the iron atom we find different shifts (using the same source). The explanation of these different shifts amounts to looking at ψ^2 (0) for an atom, in general, and in particular understanding how it is modified, in the solid.

Now let us look at some wave functions for electrons (Figure 2). The wave functions for the higher energy s electrons have more wiggles than for the lower, and lie further from the nucleus [1].

The part of the wave function in which we are most interested right now is the density at the nucleus, ψ^2 (0). One way of showing this is to actually plot the numbers vs $1s$, $2s$, $3s$, $4s$ on a logarithmic plot (Figure 3).

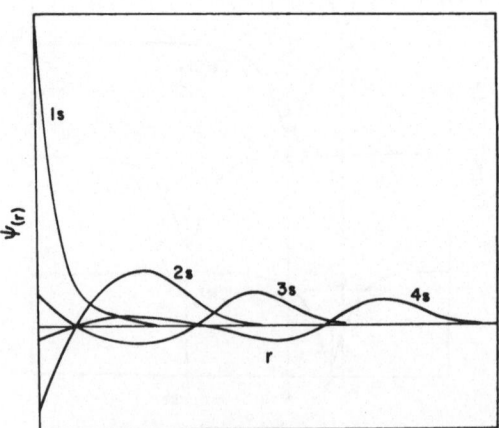

Figure 2.   Wave functions for s electrons.

In these units we find we have densities at the nucleus from each of the electrons that more or less fall along a straight line [1]. The values of $\psi^2(0)$ are roughly the same for the different ionic varieties of iron. For all practical purposes we find they are certainly the same, at least for the $1s$ plus $2s$ electrons to four or five figures. When we start getting into $3s$ electrons, we notice slight deviations, depending on the ionic state we have. Atomic iron has a full outer shell of $3s$ and $3p$ electrons plus six of $3d$, and in addition there are two of $4s$. If we worried just about the $3s$ electron density, we would get a certain value for Fe_0. Divalent iron, Fe^{2+} has a $3s^2\,3p^6\,3d^6$ outer configuration, and thus the free ion has almost the same $3s$ electron density as atomic iron. Fe^{3+} has one less electron than Fe^{2+} and that comes out of a $3d$ shell so that it has a $3s^2\,3p^6\,3d^5$ configuration. If we were to look at modern computer calculations of its $3s$ wave function, we would find that its density at the nucleus is somewhat larger than that for the Fe^{2+} configuration. The reason for the latter is that the extra d electron in Fe^{2+} increases the shielding of the $3s$ electrons from the nucleus. The density for Fe^{1+} $(3d^7)$ is even smaller for the same reason. One notes that we have no $4s$ electrons on the ions, but of course we do in metallic iron. We would find that we have some more spreading out of our $4s$ electron density depending on which configuration, such as $3d^7\,4s$ or $3d^6\,4s^2$, etc., we decided was applicable to the solid. Of course the $3d$, $4s$, and $4p$ levels are spread out and overlap each other in the solid so that in reality we have a certain distribution of each type.

Now we have all the basic building blocks we need to understand the isomer shifts. First of all, it is quite obvious that for Fe_0, metallic iron, one will have more s electrons at the nucleus than for the ions, mainly because of the $4s$ electrons. In other words, we can say $\psi^2_0(0)$ for Fe_0 has to be

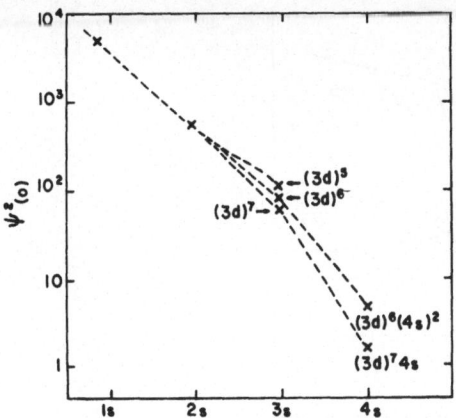

Figure 3. $\psi^2(0)$ for s electrons in iron configurations.

greater than the density from the Fe^{3+} state, but that in turn is going to be greater than that from Fe^{2+}. Therefore, from the experiments, if we are quite careful of our signs we find that

$$\delta_0 < \delta_{3+} < \delta_{2+} \tag{4}$$

but

$$\psi_0^2(0) > \psi_{3+}^2(0) > \psi_{2+}^2(0)$$

We can make everything consistent if we choose the sign of α[Eq. (3)], and therefore of $R_{ex}^2 - R_{gnd}^2$, negative. This difference can be either positive or negative, determined as a result of an atomic, solid-state, or chemical calculation.

Once you know α, $\psi^2(0)$ can be measured in other cases, showing more or less what the charge state is. For example, most Fe^{2+} isomer shifts fall close to the same value. In practice, even for insulators, we really find isomer shifts all over the scale. You can imagine why this might be. One can have orbitals that are a mixture of $4s$ and $3d$ electrons depending on the bonding. The shifts for covalent compounds are a classic example of this and reflect $3d$ transfer away from and $4s$ transfer toward the iron ion.

As another example, let us see what measurements of the isomer shift in metallic iron looks like as a function of temperature [2] and pressure [3], forgetting about the magnetic splitting that often comes into these iron experiments. Plotting $\psi^2(0)$ for iron vs pressure, a discontinuity is observed at about 130,000 atm (Figure 4). The discontinuity occurs where iron transforms from body-centered cubic to hexagonal close-packed structure. The jump corresponds to an increase in the electron density at the nucleus. There

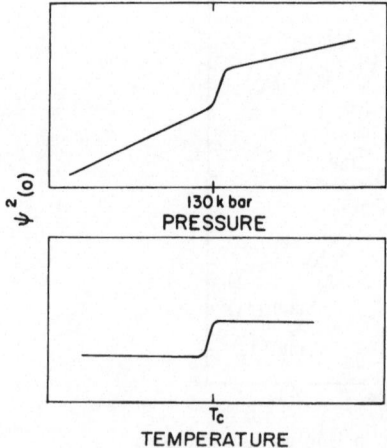

Figure 4. Isomer shift in iron *vs* pressure and temperature.

are several rather simple conclusions: isomer shifts are good for detecting phase transitions. Sometimes these are not very large anomalies but in other cases they are. Also, the direction of the change in ψ^2 (0) is interesting. Why it increases when squeezed is naively rather obvious, since when you squeeze something you expect the density to go up. This slope can be more or less explained just on the basis of volume of the atomic cell. These outer s electrons, the $4s$ electrons, are really more or less like free electrons and their density should increase as the volume decreases, and it does. The same thing goes for the discontinuity in the isomer shift. That is pretty much explained by volume change alone. The slope of the shift in the high-pressure phase still has not been explained, however, since ψ^2 (0) does not vary as strongly with the volume in this case.

If we apply temperature to metallic iron we find that the isomer shift is more or less constant, excluding the thermal shift, until the Curie temperature is reached (Figure 4). It then jumps very sharply within 0.3 degrees and goes on. This is something we don't yet understand, possibly telling us that the transition is first order.

To conclude this section, one can say that the qualitative nature of the isomer shift is generally well understood. However, the quantitative calculations are still in a relatively undeveloped state, mainly because sufficiently accurate wave functions for the solids do not yet exist!

2. MAGNETIC HYPERFINE STRUCTURE

In many Mössbauer experiments we deal with a sample in which an "effective" magnetic field H exists at the Mössbauer nucleus, permitting us to observe Zeeman splitting of the recoilless gamma ray. Although it is possible to create large enough fields by external means to cause some splitting, the latter is caused, by and large, by the electrons on the Mössbauer atom itself. As explained in Chapter 3, the interaction $H_N = -(\mu/I)H \cdot I$ splits each nuclear level into $2I+1$ sublevels so that the Mössbauer gamma ray is divided into many components, six in the case of ^{57}Fe (Figure 5).

The hyperfine magnetic field from a single electron is [4]

$$H = -2\mu_B\left[\frac{8\pi}{3}\psi^2(0)(\mathbf{s}) + \left(\frac{1}{r^3}\right) + \left(\frac{3\mathbf{r}(\mathbf{s}\cdot\mathbf{r}) - r^2\mathbf{s}}{r^5}\right)\right] \tag{5}$$

The first term is the Fermi contact interaction and is only operable for s electrons. Its direction is easy to remember if one imagines the nucleus to be inside a spherical electron distribution whose magnetization direction is opposite to its spin (Figure 6). To calculate the strength of such fields one may take the values of ψ^2 (0) for the electrons of iron as shown in Figure 3. The field from a single $1s$ electron would be a healthy 2.8×10^9 Oe (!), with the

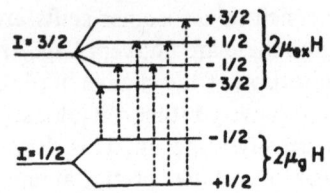

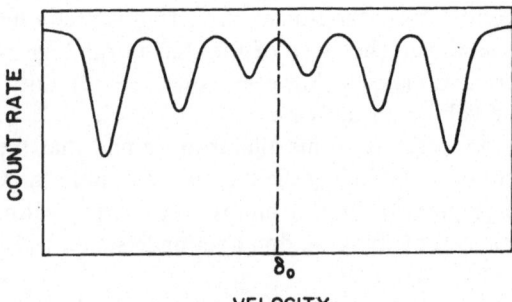

VELOCITY

Figure 5.   Magnetic hyperfine splitting of ^{57}Fe in iron.

fields from the other shells decreasing an order of magnitude, shell by shell, as does ψ^2 (0). Of course, one does not see such large fields because the inner 1s, 2s, and 3s electrons are "spin paired." Thus, in a nonmagnetic atom the main contribution would be from some unpaired outer s-state electrons. Such a behavior would occur, for instance, when Cu is placed in nickel (10^4 Oe), or Au in iron (10^6 Oe).

We observe large fields in a magnetic atom such as Fe or one of its ions. In such a case the magnetic 3d electrons polarize the inner s electrons via the exchange attraction that exists between electrons of like spin. Such a field is said to be "negative," that is, it is opposite to an externally applied field. If one remembers again that the magnetic moment of the d electrons is opposite in direction to the spin, the reason for the negative field follows in a straightforward manner (Figure 7). For a transition-metal atom like iron the magnetic 3d electrons are very effective in spin polarizing the 2s electrons. Although their overlap is even greater with the 3s electrons there is partial

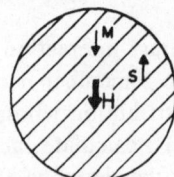

Figure 6.   Fermi contact interaction.

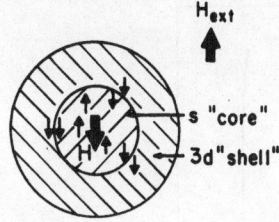

Figure 7.   Reason for a negative contact field in iron.

cancellation since the inner part of the $3s$ electron density is pulled in a direction opposite to the outer part. The net $4s$ contribution to the magnetic field is thus positive for the free iron atom, but the opposite may be true for the $4s$ conduction band in iron or its alloys. For a free iron atom the fields are about -20, -1300, $+700$, $+500$ kOe for $1s$, $2s$, $3s$, and $4s$ shells, respectively.

The second term in Eq. (5) is easily seen to be due to the orbital current. It is small in many crystals because of "quenching" of the orbital angular momentum. The third term represents the dipole field due to the electron spin. These two terms are generally smaller than the contact term but in many cases contribute appreciably. Both terms can be positive or negative depending upon the number of electrons and the shape of their charge distribution, and both terms vanish for s-state ions.

We are now in a position to understand qualitatively the measured fields at ^{57}Fe in the Fe^{2+} and Fe^{3+} ions. In the latter, $(S = 5/2, L = 0)$, the contact interaction gives about -600 kOe. In the former, Fe^{2+} $(S = 2, L = 2)$, the field is somewhat smaller because of smaller spin (hence less contact interaction), and also because of appreciable positive orbital contribution in some compounds. At room temperature the hyperfine magnetic field is about -330 kOe in metallic iron, but the various contributions still have not been clearly experimentally separated in order to compare with the various calculations that exist. Of particular importance is the question of the sign and magnitude of the contact field from the conduction electrons.

While study of the actual magnitude of hyperfine fields in general has not yet completely reached expectation, the study of their behavior under various conditions certainly has. Of particular use is the fact that the hyperfine field closely follows the magnetization in a ferromagnetic solid or the sublattice magnetization in an antiferromagnetic substance. Thus, Mössbauer experiments are valuable for studies of the temperature or pressure dependence of magnetization, as well as for detecting critical temperatures or various magnetic phases.

Again, iron represents an ideal example of the above. Let me show you some experimental results of its internal field (Figure 8). At low temperatures we of course obtain the previously described hyperfine field of -330 kOe.

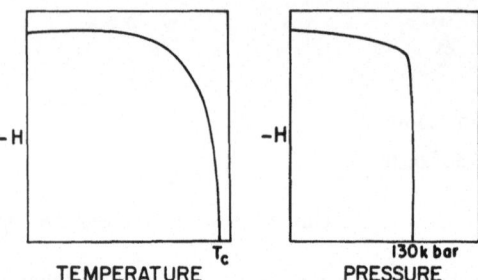

Figure 8. Hyperfine field in iron *vs* temperature and pressure.

Then as the iron is heated, we find that all the lines of the Mössbauer spectrum gradually collapse into a single line centered about the isomer-shift value we discussed earlier [2]. This occurs at the Curie temperature, which is about 1040°K. The magnetic moment of our iron atom, which is mainly from the $3d$ electrons, now flips rapidly so it is really averaging out to zero in a time which is short compared to the nuclear times involved. The nuclear levels are not split once we go into the paramagnetic state. In the ferromagnetic state the nucleus does follow and sees a field resulting from the proper time and thermal averaging. The experimental results for iron vs temperature are quite well understood.

As a function of pressure [3], the magnetic field at "zero" pressure (atmospheric pressure) is again -330 kOe and decreases slightly with pressure, going to zero at our 130 kbar phase transition (Figure 8). Thus, the body-centered cubic phase, although magnetic, goes into a hexagonal phase, which apparently is paramagnetic at room temperature. Other types of magnetic experiments are rather difficult at such high pressures. Why it is paramagnetic is still not calculated nor clearly understood.

The study of the magnetic field in the impurity Fe^{2+} ion in antiferro-

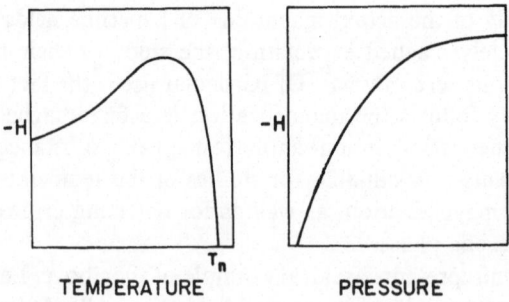

Figure 9. Hyperfine field at ^{57}Fe in CoO.

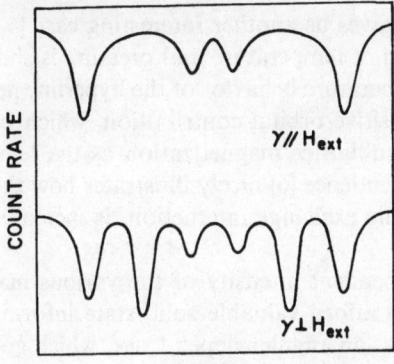

Figure 10. Angular dependence of the hyperfine splitting in iron.

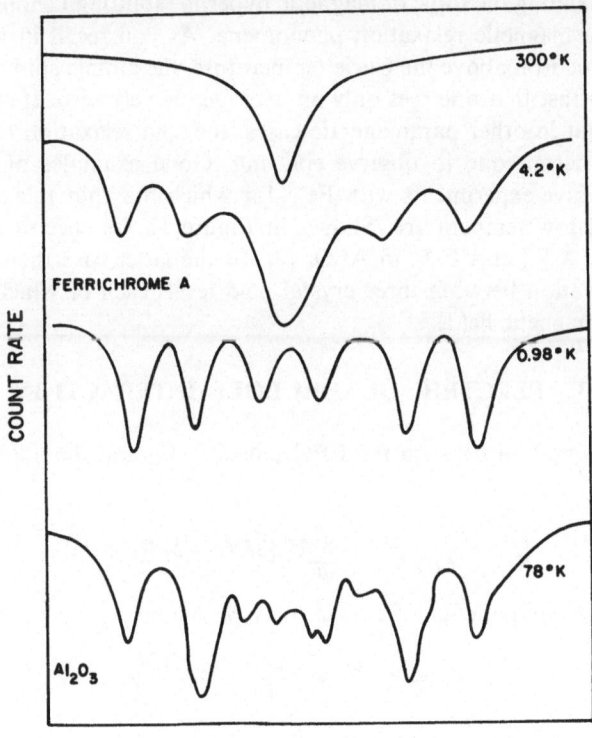

Figure 11. Relaxation effects with Fe^{3+} in ferrichrome A and corundum. Adapted from [7, 8].

magnetic CoO, also gives us another interesting case [5,6]. The behavior of the field as a function of temperature and pressure is shown in Figure 9.

The strange temperature behavior of the hyperfine field is predominantly caused by a large positive orbital contribution, which becomes increasingly unquenched by the sublattice magnetization as the temperature is lowered [5]. The pressure dependence [6] nicely illustrates how the Néel temperature, and thus the interionic exchange interaction, is increased by decreasing the lattice parameter.

The angular dependent intensity of the various magnetic split gamma ray components also afford valuable solid-state information. For ^{57}Fe the $\Delta m = 0$ transition has an angular dependence, which goes as $\sin^2\theta$ while the $\Delta m = \pm 1$ transition varies as $1 + \cos^2\theta$. Thus, for single crystals or for uniformly magnetized samples one may obtain the direction of the magnetization by simply observing the Mössbauer effect in several directions, as shown in Figure 10.

Before ending the topic of magnetic hyperfine splitting I should mention some of the magnetic relaxation phenomena. As you recall in the case of ferromagnetic iron above the Curie temperature, the atomic spin is changing direction so fast that one sees only an average, namely zero. It so happens, however, that in other paramagnetic cases, the spin relaxation time is long enough to permit one to observe splitting. Good examples of such phenomena involve experiments with Fe^{3+} for which the spin relaxation time increases at low temperature. Shown in Figure 11 are spectra of Fe^{3+} in ferrichrome A [7] and Fe^{3+} in Al_2O_3 [8]. In the latter case there is, in addition, relaxation between three crystal field levels each of which produces a different magnetic field!

3. ELECTRIC QUADRUPOLE INTERACTION

The interaction between the EFG tensor $-V_{ij}$ and the nuclear spin is given by

$$H_Q = \frac{eQ}{6I(2I-1)} \sum_{ij} V_{ij}[3I_iI_j - \delta_{ij}I(I+1)] \tag{6}$$

Any of these components, of course, may be written as

$$V_{ij} = \left\langle \frac{\rho(r)(3x_ix_j - \delta_{ij}r^2)}{r^3} \right\rangle \tag{7}$$

where $< \ >$ refers to an integral over all space, with our nucleus located at the origin and ϱ is the surrounding charge density.

For the case of ^{57}Fe such an interaction was shown to split the Mössbauer line by an amount (mm/sec)

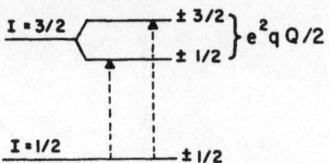

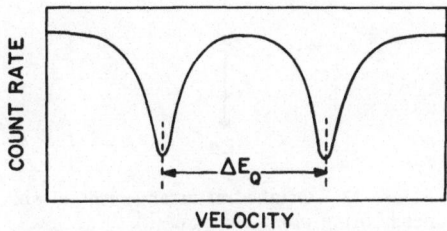

Figure 12. Quadrupole splitting of ^{57}Fe in FeSiF$_6$·6H$_2$O.

$$\Delta E_Q = \frac{1}{2}\left(\frac{c}{E_r}\right)e^2Q\sqrt{V_{zz}{}^2 + \frac{1}{3}(V_{xx} - V_{yy})^2} \qquad (8)$$

for the case where the EFG tensor is diagonal. If one is too lazy to diagonalize the EFG tensor, he can get fancy and write the square root in the symmetric yet more complicated form:

$$\sqrt{\frac{2}{9}[(V_{xx} - V_{yy})^2 + (V_{yy} - V_{zz})^2 + (V_{xx} - V_{zz})^2 + 6(V_{xy}{}^2 + V_{yz}{}^2 + V_{xz}{}^2)]} \qquad (9)$$

Equations (6) and (7) are quite general. For simplicity, here we shall merely study the component q, or $V_{zz}/e = (1/e)\langle\varrho(r)(3\cos^2\theta - 1)/r^3\rangle$, that is we assume a diagonal axially symmetric EFG tensor. Thus Eq. (6) can be shown to reduce to

$$H_Q = \frac{e^2qQ}{4I(2I - 1)}[3I_z{}^2 - I(I + 1)] \qquad (10)$$

and for ^{57}Fe, such an interaction leads to the Mössbauer spectrum shown in Figure 12.

For open shell ions (non-s state), such as Fe^{2+}, the principal source of q, namely q_{val}, comes from integrating over the ("valence") electrons on the Mössbauer ion itself, plus a secondary contribution q_{lat} from the rest of the crystal (usually assumed to be a sum over point changes). Therefore, $q =$

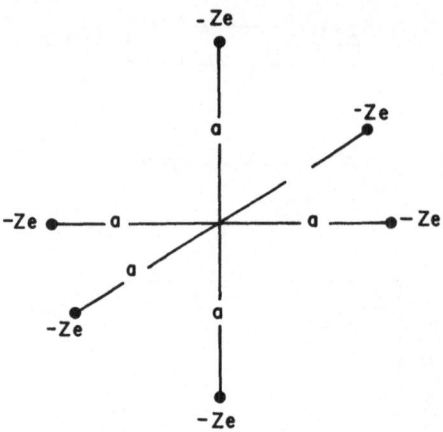

Figure 13. Octahedral arrangement of six
neighboring point charges.

$(1 - R)q_{\text{val}} + (1 - \gamma_\infty)q_{\text{lat}}$, where Sternheimer factors $-R$ and $-\gamma_\infty$ denote
the fraction of each EFG source arising from deformed closed subshells in
the Mössbauer atom. Of course both terms q_{val} and q_{lat} generally vanish
together in cubic symmetry, and for closed shell ions (s state), such as Fe^{3+},
the term $(1 - R)\, q_{\text{val}}$ always vanishes.

Let us now consider a Mössbauer ion to be at the center of a regular
octahedron formed by six nearest neighbor ions. We assume these neighbors
to be points, having charge $-Ze$ at a distance a from our nucleus (Figure
13). Such a simple octahedral arrangement occurs not only in many transition metal compounds but also in many simple salts, i.e., ordinary table
salt.

It is easily seen that a compression Δa along a tetragonal axis [100]
would yield $q_{\text{lat}} = -12Z(\Delta a/a^4)$. Whereas the same compression along the
trigonal axis [111] would yield $q_{\text{lat}} = +12\sqrt{3}\,Z(\Delta a/a^4)$. The latter type of
distortion is observed for Fe^{3+} in Fe_2O_3 and causes a small splitting ΔE_Q of
about 0.44 mm/sec ($eq \cong 10^{17}$ V/cm²).

The case of an open shell ion such as Fe^{2+} in distorted octahedral
crystal field of the above type is somewhat more complicated. In the absence
of any distortion a degenerate orbital triplet of "flowerlike" states is lowest.
Although each of the states produces a large contribution $|q_{\text{val}}| = \frac{4}{7}\langle r^{-3}\rangle$
along its respective axis, together their contributions cancel. Any deforma-
tion of the perfect symmetry will remove the degeneracy and produce a
crystal field splitting of the orbital triplet. Under the above tetragonal compression, two of the orbitals are raised in energy, such that the ground-state
orbital has a negative charge distribution that lies in a plane perpendicular

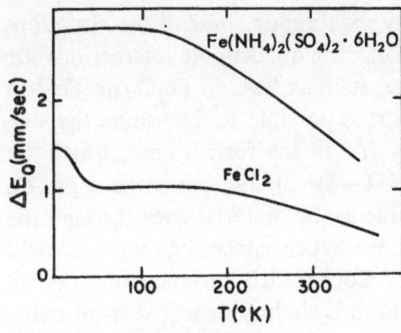

Figure 14. Temperature dependence of the quadrupole splitting of Fe^{2+} in Fe $(NH_4)_2 \cdot (SO_4)_2 \cdot 6H_2O$ and $FeCl_2$. Adapted from [9,10].

to the tetragonal axis. The latter, in turn, yields a positive value, $q_{val} = \frac{4}{7}$ $<r^{-3}>$. However, under a trigonal compression the lowest state is an orbital singlet with a negative charge distribution along the trigonal axis, so that in this case $q_{val} = -\frac{4}{7} <r^{-3}>$. Thus, for our simple point-charge octahedral complex, q_{lat} and q_{val} are seen to be opposite in sign. The quadrupole splitting for Fe^{2+} in the trigonal salt $FeSiF_6 \cdot 6H_2O$ is 3.7 mm/sec ($eq \simeq -10^{18}$ V/cm^2). The lattice contribution is usually an order of magnitude smaller than the valence contribution in ferrous compounds. In the event that the tetragonal or trigonal crystal field splitting is of the same order as kT, the higher orbital states can appreciably reduce the value of q_{val}. Such a behavior presupposes a relaxation between these states which is rapid compared to the nuclear precession frequency. Indeed, for this reason most Fe^{2+} compounds show a quadrupole splitting which decreases at high temperatures. Shown in Figure 14 is the temperature dependence for Fe $(NH_4)_2(SO_4)_2 \cdot 6H_2O$ [9] and $FeCl_2$ [10]. The latter shows the additional contribution caused by low-temperature magnetic exchange interaction, which will be discussed below. If, however, the relaxation between the crystal field states is slow, one expects no reduction since each state produces the same splitting. An interesting example of this type is Fe^{2+} in MgO [11]. Although this material is cubic the relaxation between the lowest strain-split spin-orbit triplet is sufficiently slow below 20°K to permit observation of a small splitting (0.33 mm/sec).

Now that we have described the EFG tensor in connection with several solid-state applications it is fair to ask, "what are the problems?" Unfortunately, just as in the isomer-shift and magnetic hyperfine splitting cases, the trouble with the quadrupole splitting is mainly quantitative. One is forever asking himself how does the solid modify the wave functions that were so precisely determined for the free atom or ion. This question not only concerns the calculations of q_{val} and q_{lat}, but the Sternheimer corrections as well. It appears that at the present time the best one can hope for is perhaps 15 to 25% accuracy.

I will conclude this section by briefly mentioning some of the situations that can occur if magnetic hyperfine and electric quadrupole interactions are present simultaneously. In the usual case, such as Fe^{3+} in Fe_2O_3 or Fe^{2+} in CoO [5], both fields are static, and then it is possible to determine the sign of the component of V_{ii} which is along H.[2] In the former case, which we recall has axial symmetry $[V_{xx} = V_{yy} = (-\frac{1}{2})V_{zz}]$, the component of V_{ii} along H appears to change sign and double as the material goes through the Morin transition. But this is just what we expect since the magnetic field changes its direction by 90°. The case of CoO is also an interesting one in many respects. For example, this compound is slightly distorted from cubic symmetry below its Néel temperature of 292°K (see Figure 9), yet gives substantial EFG tensor, one much greater than that in MgO. In this case an appreciable fraction of the quadrupole splitting comes as a consequence of the sublattice magnetization, which, having a particular direction associated with it, can alone destroy the cubic symmetry about the nucleus. The same is true for $FeCl_2$, discussed above.

4. LATTICE DYNAMICS

In Chapter 1 you learned that for harmonic solids, the recoil-free fraction f permits a measurement of the mean-square displacement $<x^2>$ of the Mössbauer atom or ion. In general, f measurements and their pressure and temperature dependence can provide good checks on lattice dynamical models, whether they apply to an impurity or a host atom.

An interesting complication comes into effect for compounds in which the binding of the Mössbauer atom is anisotropic, so that f varies with the direction of gamma-ray emission or absorption. In such an instance the intensities of the quadrupolar split gamma ray can reflect the anisotropy, even for a powder [12]. In the absence of anisotropy for ^{57}Fe in an axially symmetric EFG, the $\pm\frac{1}{2}\rightarrow\pm\frac{3}{2}$ transition has an angular dependence of 3 $(1+\cos^2\theta)$, whereas the $\pm\frac{1}{2}\rightarrow\mp\frac{1}{2}$, $\pm\frac{1}{2}$ transition has an angular dependence of $5-3\cos^2\theta$. For single crystals the ratio is thus three along the z direction but 3/5 perpendicular to the z axis. An anisotropic f factor will, in general, change these ratios slightly. For an isotropic powder this ratio is unity but in some cases anisotropy can appreciably alter it.

Another interesting feature is anharmonicity, that is, the vibrating atom that sits in a potential well which is not quite harmonic. This is easy to imagine if we place our emitting atom between two neighbors both of which produce identical attractions (Figure 15). Thus, in cases where the Mössbauer atom is in a large cage of atoms the binding is expected to be

---

[2] One also may determine the sign of q by applying an external field.

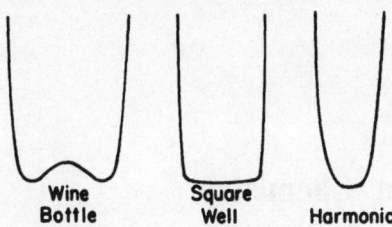

Wine Bottle Square Well Harmonic

Figure 15. Dependence of a potential well upon near neighbor distance.

anharmonic. This results in anomalously large mean-square displacements (and also low f factors, especially at low temperatures). A good example of this is a comparison of f factors for Fe^{2+} in FeF_2 and $FeCl_2$ [13]. Because of more "rattle space," anharmonicity is found in the latter case but not in the former.

In principle, all the above statements apply to the thermal shift as well. That is, $<v^2>$ is another quantity that emerges from a theory of lattice dynamics. Moreover, it too is affected by such things as temperature (of course), pressure, anisotropy, anharmonicity, etc. The main problem, however, lies in the fact that these shifts are generally small and rather insensitive to attempts to make them vary.

REFERENCES

1. R. E. Watson, Solid State and Molecular Theory Group, *M.I.T. Tech. Rept.* No. 12 (1959).
2. R. S. Preston, *Phys. Rev. Letters* 19, 75 (1967).
3. D. N. Pipkorn, C. K. Edge, P. Debrunner, G. DePasquali, H. G. Drickamer, and H. Frauenfelder, *Phys. Rev.* 135, A1604 (1964).
4. R. E. Watson and A. J. Freeman, *Phys. Rev.* 123, 2027 (1961).
5. N. N. Ok and J. G. Mullen, *Phys. Rev.* 168, 563 (1968).
6. C. J. Coston, R. Ingalls, and H. G. Drickamer, *Phys. Rev.* 145, 409 (1966).
7. H. H. Wickman, M. P. Klein, and D. A. Shirley, *Phys. Rev.* 152, 345 (1966).
8. G. K. Wertheim and J. P. Remeika, *Phys. Letters* 10, 14 (1964).
9. R. Ingalls, K. Ono, and L. Chandler, *Phys. Rev.* 172, 295 (168).
10. K. Ono, A. Ito, and T. Fujita, *J. Phys. Soc. Japan* 19, 2119 (1964).
11. H. R. Leider and D. N. Pipkorn, *Phys. Rev.* 165, 494 (1968).
12. S. V. Karyagin, *Dokl. Akad. Nauk SSSR* 148, 1102 (1963) (in Russian); *Proc. Acad. Sci. USSR, Phys. Chem. Sect.* 148, 110 (1964) (in English).
13. D. P. Johnson and J. G. Dash, *Phys. Rev.* 172, 985 (1968).

Chapter 6

Application to Coordination Chemistry

J. Danon

Departamento Fisica Molecular e Estados Solido
Centro Brasileiro de Pesquisas Fisicas
Rio de Janeiro, Brasil

There are a number of review papers discussing the different aspects of the Mössbauer effect in connection with chemical concepts [1–4]. In the present chapter we shall take the point of view of a coordination chemist who asks to what extent the Mössbauer effect can be of any help in the specific problems of his field.

Coordination chemistry deals with coordination or complex compounds. Such compounds contain a central ion M bonded to several ligands L, L', L'', etc. Although complexes can be formed by all electropositive elements, M is, in most cases, a transition-metal ion.[1] We are in a fortunate position here since one of the most important transition elements is iron, for which ^{57}Fe is available, which is the most favorable isotope for Mössbauer spectroscopy. Besides iron, Mössbauer investigations have been made with coordination compounds of Ni, Ru, Os, Ir, Pt, Au, Eu, and Np among the transition elements, and of Sn, Sb, Te, I, Xe, and Kr among the nontransition elements.

Typical problems of coordination chemistry frequently deal with the question of the number and arrangement of the ligands in a complex compound.

To a given stoichiometric formula ML' L . . .", we can have a variety of geometrical dispositions of the ligands in the coordination sphere. For example, the complex $Co(CN)_5H_2O$ is penta- or hexacoordinated according to whether the water molecule is or is not bonded to the Co ion (Figure 1). If this complex is pentacoordinated, it can be a trigonal bipyramid or a

---

[1] An illuminating discussion of the concept of complex compounds and their evolution in the development of modern inorganic chemistry is given in Cotton and Wilkinson's book [5].

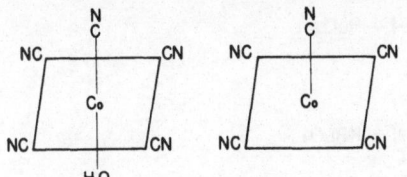

Figure 1. Hexa- and pentacoordinated structures for complex cobalt (II) cyanide.

square pyramid. In any of these geometrical arrangements we may have a complex where all the CN are bonded to the Co through the carbon, or some through the N end of the CN ligand, illustrating a case of the so-called ligand isomerism.

Other coordination chemistry problems concern the nature of chemical bonding in the complex. The simplest and at the same time very fundamental question is that of the valence state of the central metal ion in the complex. A further step is to consider the electronic structure of the central ion, taking into account the perturbation introduced by the electric field from the ligands. The aim of this ligand field approach is to determine the electronic wave function for the ground state of the central ion. In a more elaborate approach, such as that of the molecular orbital theory, the purpose is to establish the energy levels and the charge distribution at the central ion and at the ligands of the complex ion.

A feature common to all spectroscopic methods is that they can be used in two ways: on the basis of selection rules and symmetry arguments and on the basis of a detailed analysis involving the electronic structure of the molecule. Using the first procedure we derive data on molecular architecture, valency states, nature of the ligands, etc., without any detailed theoretical analysis regarding the molecule. Typical examples are given by infrared and Raman spectroscopies from which practically all data are obtained though the assignment of the absorption bands, which are classified according to selection rules and symmetry arguments. However, it is also possible to make a detailed analysis of force constants on the basis of the electronic structure of the molecule. On the other hand, in a spectroscopic method, such as electron spin resonance, the interpretation of most spectra requires the location of the unpaired electron in a given electronic energy level. For this reason in this spectroscopic method data are derived mainly from the analysis of the electronic structure of the molecule.

Mössbauer spectroscopy also has a dual character. The hyperfine interactions on which Mössbauer spectroscopy is based are subject to selection rules and symmetry arguments. The basic selection rule assigns distinct ranges of values of the isomer shift for different oxidation states of an element. Using this selection rule we are able to classify the complex compounds of an element according to the oxidation state of the central ion.

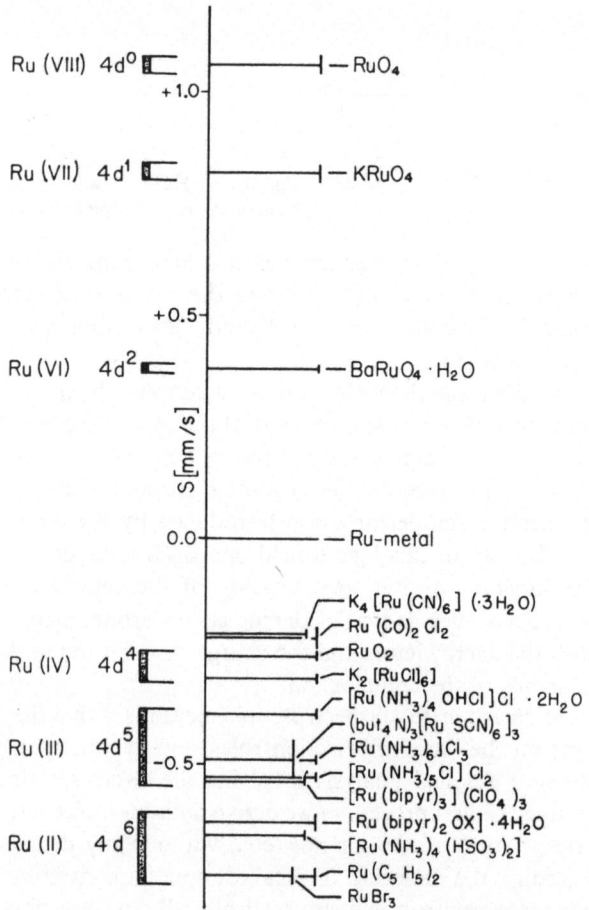

Figure 2. Graphical representation of the isomer shifts
obtained for various ruthenium compounds.

Other hyperfine interactions, such as quadrupole coupling and magnetic
splitting, are also functions of the oxidation state of the element, but in a
less general way as compared to the isomer shift. Figure 2 illustrates the use
of the isomer-shift selection rule in ruthenium coordination chemistry [6].

The basic symmetry argument of Mössbauer spectroscopy involves
nuclear quadrupole coupling and explains the absence of quadrupole
splitting when the surroundings of the nucleus have cubic symmetry. By this
symmetry argument we can determine whether the molecular geometry
around an element is or is not distorted from octahedral symmetry. More-

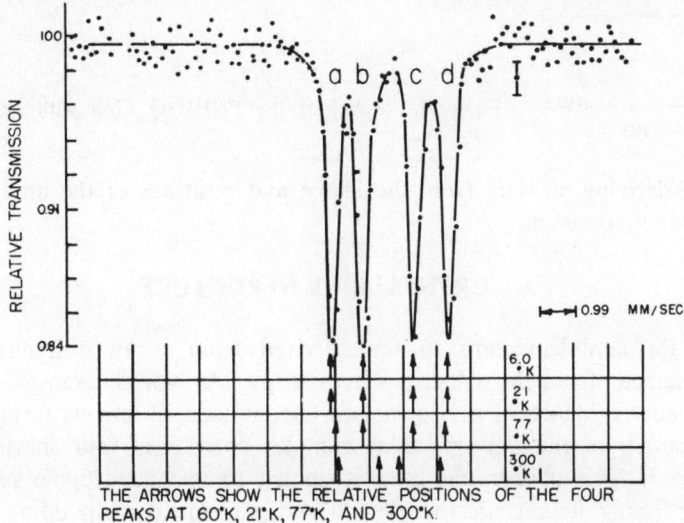

THE ARROWS SHOW THE RELATIVE POSITIONS OF THE FOUR
PEAKS AT 60°K, 21°K, 77°K, AND 300°K.

Figure 3. Mössbauer spectrum of ferrous formate at liquid helium
temperature. The four peaks are labeled a, b, c, and d. Adapted from [7].

over, the value of the quadrupole coupling is sensitive to the *symmetry* of
the electronic distribution around the nucleus. Thus, in the case of iron the
range of values of the quadrupole couplings are distinct for the different
spin configurations of a given oxidation state. On this basis it is frequently
possible to decide between the high- or low-spin configuration for the iron
ion in complex compounds.

We shall now illustrate the applications of Mössbauer spectroscopy to
coordination chemistry based on these selection rules and symmetry argu-

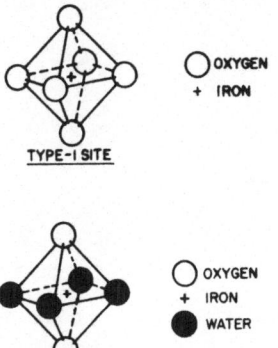

Figure 4. Schematic of the approximate nearest-
neighbor symmetry at the two ferrous ion sites in ferrous
formate.

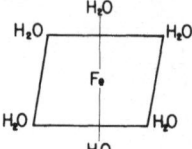

Figure 5. Model of the $Fe(H_2O)_6^{3+}$ ion with octahedral structure.

ments, deriving all data from the shape and positions of the lines in the Mössbauer spectrum.

1. CRYSTALLINE STRUCTURE

Data complementary to that derived from x-ray diffraction can be obtained from Mössbauer spectroscopy. A typical example is the Mössbauer evidence of nonequivalent ferrous ions in ferrous formate [7]. The spectra of the polycrystalline complex consists of four sharp peaks (Figure 3). This pattern has been attributed to two quadrupole splittings arising from different effective electric field gradients corresponding to two nonequivalent ferrous ion sites. Figure 4 shows the nearest neighbor symmetry of the two sites: one is O_h and the other is D_{4h}.

2. COMPLEX ISOMERISM

Compounds with the same stoichiometric composition but different arrangement of the ligands are called isomers. The Mössbauer spectrum

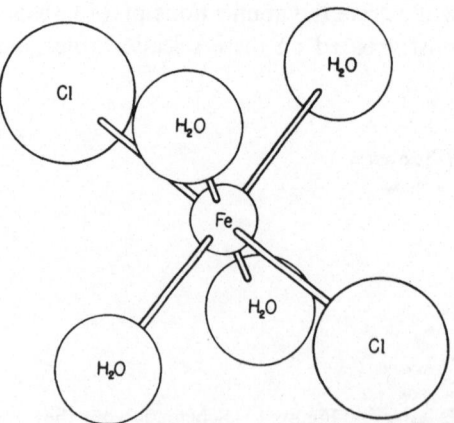

Figure 6. Model of the $(FeCl_2(H_2O)_4)^+$ ion.

Cl
| NH₃
| ⟋
Cl——Co——NH₃
⟋|
NH₃ |
    NH₃

cis

NH₃
| NH₃
| ⟋
Cl——Co——Cl
⟋|
NH₃ |
    NH₃

trans

Figure 7. *Cis* and *trans* isomers for the octahedral structures of $Co(NH_3)_4Cl_2$.

may be of great help for deciding the different possibilities for the structure of an isomer.

The common hydrated ferric chloride $FeCl_3 \cdot 6H_2O$ was assumed to have the iron ion surrounded by an octahedral environment of water molecules as shown in Figure 5. But the Mössbauer spectrum of this molecule [8] exibits much quadrupole splitting, and on the basis of this result it was suggested that the symmetry around the ferric ion should be lower than octahedral. This was the starting point of x-ray diffraction investigations [9] which have shown that the correct structure is that of an hydrated isomer, as is illustrated in Figure 6. This result is of interest in the understanding of the properties of aqueous solutions of the ferric ion in the presence of chloride ions, since the different complexes which are formed are derived from this basic distorted structure [10].

2.1. *Cis-Trans* Isomerism

This frequent case of isomerism is illustrated in Figure 7. The differences in ligand disposition should induce different values of the electric field gradient at the central ion: in the *trans* case there is axial symmetry (D_{4h}), which is absent in the *cis* case (C_{2v}). In a treatment based on a point-charge model the field gradient is given by

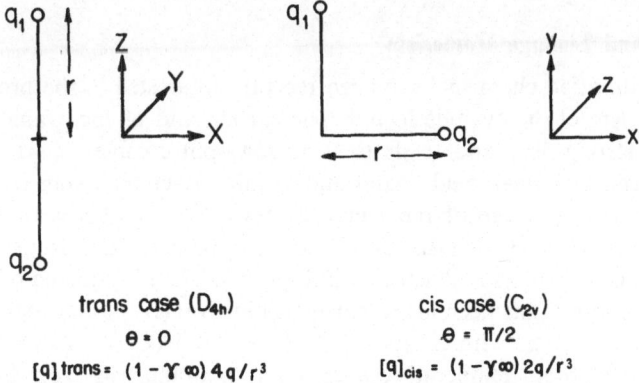

trans case (D_{4h})

$\theta = 0$

$[q]_{trans} = (1 - \gamma_\infty) 4q/r^3$

cis case (C_{2v})

$\theta = \pi/2$

$[q]_{cis} = (1 - \gamma_\infty) 2q/r^3$

Figure 8. The electric field gradient in a point-charge model for *cis* and *trans* isomers.

Table 1. Mössbauer Hyperfine Parameters of *Cis–Trans* Isomers[a]

Absorber	ΔE_Q, mm/sec	δ, mm/sec
1. (Fe(CNMe)$_6$)(HSO$_4$)	0.00	0.16
2. *cis*-Fe(CNMe)$_4$(CN)$_2$	0.24	0.18
3. *trans*-Fe(CNMe)$_4$(CN)$_2$	0.44	0.18
4. (Fe(CNEt)$_6$)(ClO$_4$)$_2$	0.00	0.18
5. (Fe(CN)(CNEt)$_5$)(ClO$_4$)	0.17	0.22
6. *cis*-Fe(CNEt)$_4$(CN)$_2$	0.29	0.23
7. *trans*-Fe(CNEt)$_4$(CN)$_2$	0.59	0.23
8. (Fe(CNCH$_2$Ph)$_6$)(ClO$_4$)$_2$	0.00	0.14
9. Fe(CN)(CNCH$_2$Ph)$_5$ClO$_4$	0.28	0.16
10. *trans*-(Fe(CN)$_2$(CNCH$_2$Ph)$_4$)	0.56	0.17
11. *cis*-Fe(phen)$_2$(CN)$_2$	0.58	0.45
12. "*trans*"-Fe(phen)$_2$(CN)$_2$	0.60	0.51

[a] Errors for δ and $\Delta E_Q = \pm 0.05$ mm/sec; δ values are relative to sodium nitroprusside.

$$q = (1 - \gamma_\infty)\{\sum_i q_i(3\cos^2\theta_i - 1)/r_i^3\},$$

where $(1-\gamma_\infty)$ corrects for antishielding effects and q_i is the magnitude of the ith charge, whose coordinates are r_i and θ_i (see Chapter 4, p. 77). Since $\theta = 0$ for the *trans* case and $\theta = \pi/2$ for the *cis* case, as is illustrated in Figure 8, one finds that the ratio of quadrupole splittings is $\Delta E_{Q(trans)}/\Delta E_{Q(cis)}) = 2:1$. Table 1 lists the results obtained [11] in a series of low-spin *cis–trans* isomers of iron (II). Within the experimental error the ratio of 2:1 is observed. The results suggest that the *cis–trans* isomerism reported for dicyanobis-(1,10-phenantroline) iron (II) is doubtful. One can identify by this simple procedure which is the *cis* and which is the *trans* isomer.

2.2. Ligand Linkage Isomerism

Coordination chemists have been recently interested in the problem of the isomerism of the cyanide ligand. The carbon end of the cyanide ligand creates a strong field and tends to form low-spin complexes, whereas the nitrogen end is a weak field ligand and usually gives high-spin complexes.

The x-ray structure of the metal ferrocyanides (the Prussian blue and similar compounds) shows that one metal is bonded to the nitrogen and the other to the carbon, as is illustrated in Figure 9. The phenomena of cyanide linkage isomerism has been investigated [12] in iron (II) hexacyanochromate (III). This complex, with composition $Fe_3^{II}(Cr^{III}(CN)_6)$, exibits linkage isomerism: at room temperature it changes spontaneously from form I to form II, as illustrated in Figure 10.

According to the arguments previously discussed concerning the difference in strengh of the ligand field induced by both ends of the CN

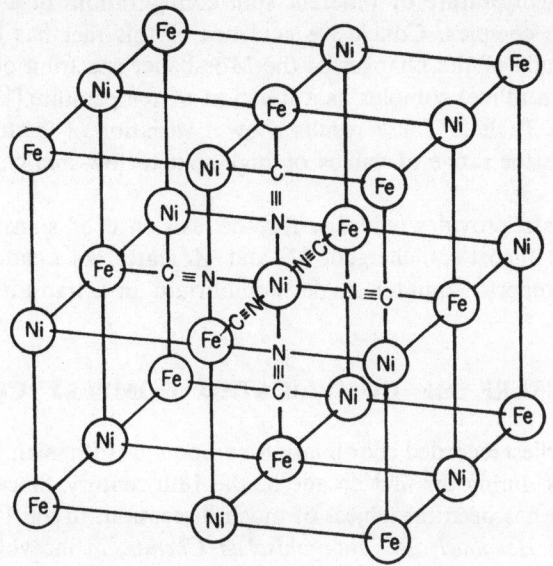

Figure 9. The structure of $K_2NiFe(CN)_6$.

$F_e^{+2} - N \equiv C - C_r^{+3}$

$C_r^{+3} - N \equiv C - F_e^{+3}$ Figure 10. Ligand linkage isomers in iron hexacyanochromate.

ligand, one should expect a high-spin complex of Fe^{2+} in form I and a low-spin iron (II) complex in form II. The Mössbauer spectra of the form I shows indeed the typical isomer shift and quadrupole splitting of high-spin Fe^{2+}. As the spontaneous isomerization process leads to form II, the quadrupole coupling decreases going to the range of low values characteristic of the ferrous ion in the low-spin configuration.

2.3. Spin-State Equilibria

Magnetic susceptibility measurements have suggested the coexistence

Table 2. Mössbauer Hyperfine Parameters of Iron (II)-bis-(1,10 phenantroline)

T, °K	ΔE_Q, mm/sec	δ^a, mm/sec
293	2.67 ± 0.03	0.43 ± 0.03
77	0.34 ± 0.06	0.62 ± 0.05

^{a} δ is relative to sodium nitroprusside.

at a given temperature of different spin configurations of a transition ion in the same complex. Conclusive evidence of this fact has been obtained from the study of the changes of the Mössbauer spectrum of iron (II)-bis-(1,10-phenantroline) complex as a function of temperature [13]. The results are listed in Table 2. These results show a variation of the hyperfine parameters from the range of values of high-spin to low-spin configuration of the ferrous ion.

This data provides evidence for the existence of spinstate equilibria between the almost equienergetic $^5T_2$ and $^1A_1$ states. As a matter of fact this is the first reported quintet–singlet equilibrium in a transition-metal complex.

3. STRUCTURE OF COMPLICATED COMPLEX COMPOUNDS

The earliest recorded coordination compound is Prussian blue, obtained by Diesback during the first decade of the 18th century. Since its discovery its structure has been the object of much discussion. In the 1930 edition of the *Gmelins Handbuch der Anorganischen Chemie*, in the volume on iron, more than a thousand pages are devoted to the properties of this and related complexes.

The following are the main questions regarding Prussian blue, both in the soluble and insoluble form, and Turnbull's blue.
1. Which are the electronic configurations of the two kinds of iron in these compounds: ferric ferrocyanide Fe^{III} (Fe^{II} $(CN)_6$), ferrous ferricyanide Fe^{II} (Fe^{III} $(CN)_6$), or a more complex configuration?
2. Are Prussian blue, made by mixing the solutions of ferric compound and ferrocyanide ion, and Turnbull's blue, made by mixing the solutions of ferrous compound and ferrocyanide ion, the same compound or not?

We find the answer to these questions, which have been the object of many discussions, in a recent report by A. Ito *et al.* [14] "Mössbauer Study of Soluble Prussian Blue, Insoluble Prussian Blue and Turnbull's Blue" in 13 pages.

The Mössbauer spectra were observed in the range of 1.6–300°K. At the lowest temperature the spectra are well resolved and show a superposition of hyperfine split levels and a single line for all the compounds, as is illustrated in Figure 11. The results obtained are listed in Tables 3 and 5 and compared with typical values of Mössbauer parameters for the various states of iron (Table 4).

By comparing the internal fields for the two kinds of iron ions with typical values, it is seen that $H_n = 540$ kOe is just the value for high-spin Fe^{3+}, and a value of zero corresponds to low-spin Fe^{2+}. The isomer shift and

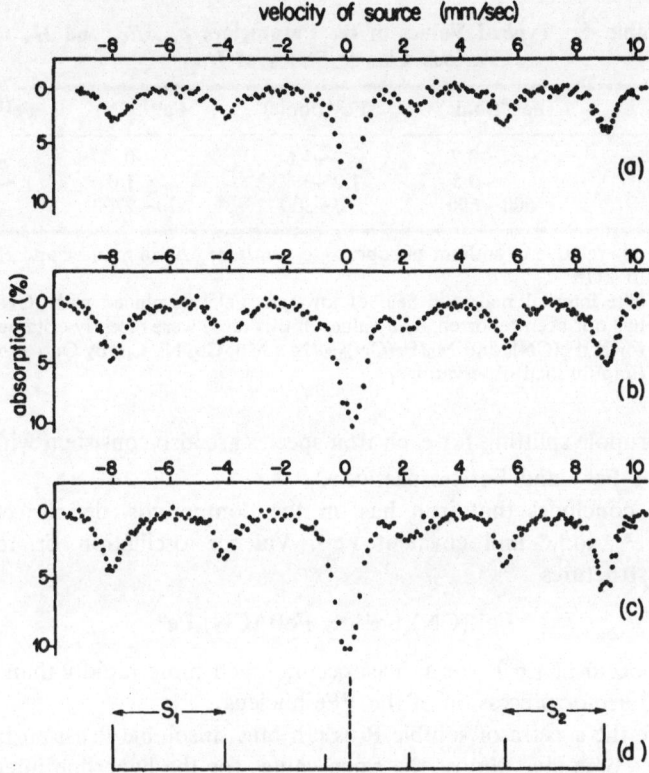

Figure 11. The Mössbauer spectra obtained at 1.6°K for (a) soluble Prussian blue, (b) insoluble Prussian blue, and (c) Turnbull's blue. The line positions for the two kinds of iron are indicated in (d) with solid lines and dashed line.

Table 3. Values of the Parameters δ, $(S_1\text{-}S_2)$, and H_n[a] for Prussian Blue and Turnbull's Blue

| Parameter | Prussian blue | | | | Turnbull's blue | |
| | soluble | | insoluble | | | |
	Fe^{3+}	Fe^{II}	Fe^{3+}	Fe^{II}	Fe^{3+}	Fe^{II}
δ	0.84 ± 0.05	0.33 ± 0.05	0.84 ± 0.05	0.31 ± 0.05	0.83 ± 0.05	0.27 ± 0.05
$S_1\text{-}S_2$	0.37 ± 0.15	0	0.48 ± 0.15	0	0.52 ± 0.15	0
H_n	536 ± 20	0 ± 10	541 ± 20	0 ± 10	543 ± 20	0 ± 10

[a] δ is relative to sodium nitroprusside, mm/sec; S_1 and S_2 are shown in Figure 11 and $S_1\text{-}S_2 = 1/2\ e^2qQ\ (3\cos^2\theta - 1)$.

Table 4. **Typical Values of the Parameters δ, ΔE_Q, and H_n for Various Charge States of Iron[a]**

	Fe^{3+}(ionic)	Fe^{2+}(ionic)	Fe^{III}(CN)	Fe^{II}(CN)
δ	~0.7	~1.6	~0	~0
ΔE_Q	~0.5	1.0—3.4	≤ 1.0	~0
H_n	500~600	0~300	170~270[b]	0

[a] δ is relative to sodium nitroprusside, mm/sec; ΔE_Q in mm/sec and H_n in kOe.

[b] The internal magnetic field of low-spin Fe^{III} combined with (CN)$_6$ has not been reported. The values in this table were recently obtained for $K_3Fe(CN)_6$ and $M_3(Fe(CN)_6)_2$ (M : Mn, Co, Ni, Cu) by Ono et al. (unpublished observations).

the quadrupole splitting for each iron species are also consistent with typical values for Fe^{3+} and Fe^{II} respectively.

One concludes that iron has, in the compounds, definite electronic states, Fe^{3+} ionic and covalent Fe^{II}. Valence oscillation or resonance between structures

$$Fe^{II}(CN)_6 Fe^{3+} \leftrightarrows Fe^{III}(CN)_6 Fe^{2+}$$

does not occur at 1.6°K, or at least occurs much more rapidly than 10^{-8} sec (time of Larmor precession of the ^{57}Fe nucleus).

Since the spectra of soluble Prussian blue, insoluble Prussian blue, and Turnbull's blue give almost the same values for the hyperfine interactions, we conclude that the electronic structure of these compounds is the same from a Mössbauer spectral point of view. Therefore, in Turnbull's blue, which is made by ferrous compounds and ferricyanide, the charge transfer from Fe^{2+} (high-spin) to Fe^{III} (low-spin) or flipping of the CN ligand by 180° should occur at the instant of combination.

The intensity ratios of the Mössbauer spectra are consistent with the stoichiometric formula $KFe(Fe(CN)_6)$ for soluble Prussian blue and

Table 5. **Intensity Ratio between the Two Kinds of Iron, Fe^{3+} and Fe^{II}, for Prussian Blue and Turnbull's Blue**

	Fe^{3+}/Fe^{II} observed[a]	Fe^{3+}/Fe^{II} normalized[b]
Soluble Prussian blue	1.39	1.00
Insoluble Prussian blue	1.78	1.28
Turnbull's blue	1.84	1.32

[a] Fe^{3+}/Fe^{II} observed with the absorbers containing iron of about 20 mg/cm^2.

[b] Normalized to the thin absorber.

Fe_4 (Fe $(CN)_6$) for insoluble Prussian blue and Turnbull's blue. Thus, from the comparison of the Mössbauer spectra, we obtain all the basic information on the structure of these complicated complexes.

4. ELECTRONIC STRUCTURE OF MOLECULES

Let us now demonstrate the use of Mössbauer spectroscopy for obtaining information on the electronic structure of molecules. As mentioned earlier, a theoretical model is now required in order to interpret the results.

4.1. Using Ligand Field Theory

Ligand field theory has been used for interpreting the large quadrupole splitting observed in high-spin ferrous compounds, which are temperature-dependent and vary markedly from compound to compound [15] (Figure 12). In the ferrous ion the $3d^6$ electrons are distributed in the high-spin configuration with maximum multiplicity along the five d orbitals. As is shown in Figure 13, this configuration leads to an extra electron with the spin antiparallel to the other five.

The main contribution to the field gradient at the iron nucleus is given by this extra electron. Since the configuration of five parallel spins results in spherical symmetry, their electronic contributions to the electric field gradient at the nucleus vanish. The values of the electric field gradient produced by the different $3d$ wave functions in the presence of the crystal field are listed in Table 6.

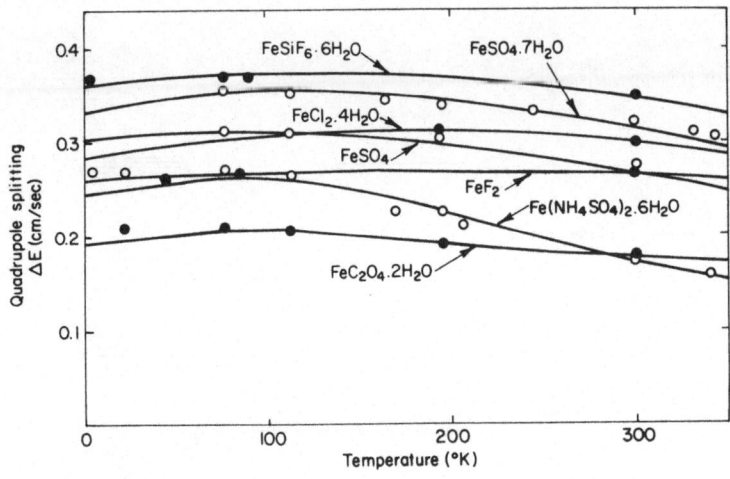

Figure 12. Quadrupole splitting of several ferrous compounds as a function of temperature. Adapted from Ingalls [15].

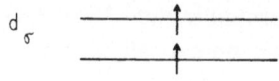

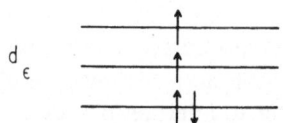

Figure 13. Spin configuration of the $3d$ electrons of ferrous ion in the high-spin case.

The orbital in which the extra electron will be placed depends on the deviation from cubic symmetry of the ligand field. As illustrated in Figure 14, axial and rhombic fields lift the degeneracy within the d_γ and d_ε shells, and further splitting of the energy levels occurs by spin-orbit coupling. The temperature-dependence of the quadrupole splitting is due to the distribution of the electron among these sublevels. Covalency effects are introduced by considering that the bonding delocalizes the $3d$ electron. A covalency coefficient $a^2 < 1$ accounts for the expansion of the radial part of the $3d$ wave function. Using this treatment it has been possible to obtain reasonable estimates of the energy splitting of the ligand field and covalency parameters by an adequate choice of the ground-state wave function in order to fit the temperature-dependence of the quadrupole splitting shown in Figure 12.

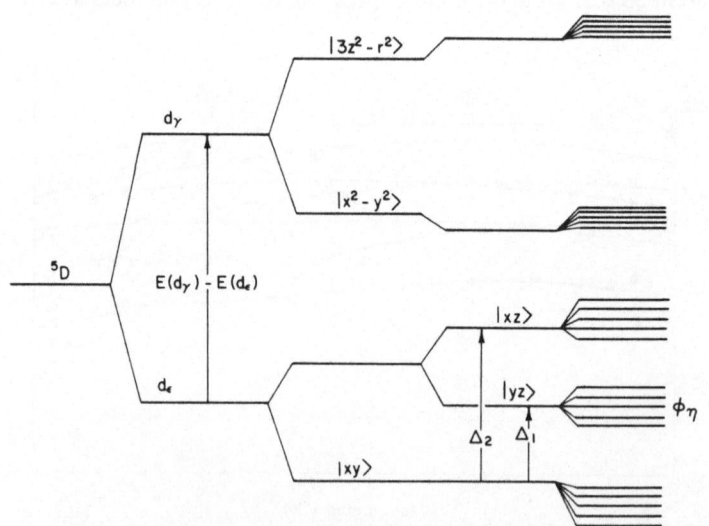

Figure 14. Energy level scheme for the ferrous ion under the action of the crystalline field plus spin-orbit coupling. Adapted from Ingalls [15].

Table 6. Values of Field Gradient q and Asymmetry Parameter η for $3d$ Electrons

Orbital		q	η
d_γ	$d_{x^2-y^2}$	$+4/7 <r^{-3}>$	0
	d_{z^2}	$-4/7 <r^{-3}>$	0
d_ϵ	d_{xy}	$+4/7 <r^{-3}>$	0
	d_{xz}	$-2/7 <r^{-3}>$	$+3$
	d_{yz}	$-2/7 <r^{-3}>$	-3

4.2. Using Molecular Orbital Theory

The use of molecular orbitals for interpreting the Mössbauer hyperfine parameters has been demonstrated [16] in the case of the nitroprusside complex $Fe^{II}(CN)_5NO^{2-}$. The molecular orbital level scheme proposed for this complex [17] is reproduced in Figure 15. The Φ wave functions can be regarded as the antibonding combinations that would be formed with a symmetric octahedron of CN ligands. They are perturbed by the ligand field distortion arising from the substitution by NO for one CN. The π^* and σ^* orbitals of NO overlap with the central ion wave functions, forming relative bonding and antibonding combinations. ψ_{xy} and all levels below are filled with paired electrons. However, the electron delocalization will be different is these orbitals since ψ_{xy} is essentially nonbonding whereas the lower doublet (ψ_{xz}, ψ_{yz}) forms strong π^* antibonds with the empty orbitals of NO. These

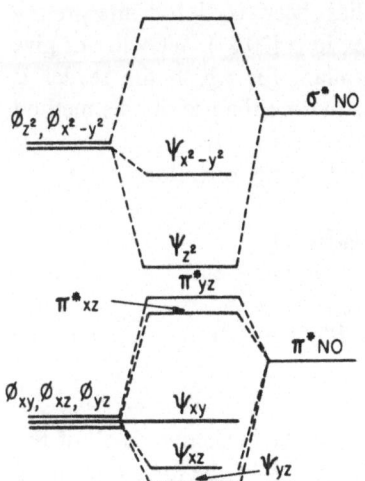

Figure 15. Energy level diagram for the nitroprusside ion.

three full levels give rise to an asymmetrical charge distribution and induce an electric field gradient at the iron nucleus which can be written as

$$q = \{(4/7)n_{xy}^2 - (2/7)(n_{xz}^2 + n_{yz}^2)\} <r^{-3}> \tag{1}$$

where the n^2 is the effective d electron population in the corresponding molecular orbital.

The quadrupole splitting of 3.60 mm/sec in a high-spin Fe^{2+} complex, such as $FeSiF_6 \cdot 6H_2O$, is due to a single $3d$ electron with field gradient

$$q = (4/7)n^2 <r^{-3}>$$

where $n^2 = 0.80$.

Taking the ratio of quadrupole splittings as the ratio of EFG, one has:

$$\frac{\Delta E_Q}{3.60} = \frac{(4/7)n_{xy}^2 - (2/7)(n_{xz}^2 + n_{yz}^2)}{(4/7) \times 0.8} \tag{2}$$

Using the calculated values [17] $n_{xy}^2 = 2(0.81)$ and $n_{xz}^2 = n_{yz}^2 = 2(0.61)$ one finds 1.8 mm/sec, which is in good agreement with the experimental quadrupole splitting of 1.726 mm/sec reported for sodium nitroprusside.

This result confirms the strong delocalization of the d_{xz}, d_{yz} iron electrons in the pentacyanonitrosyls, showing the basic importance of back-donation in determining the energy levels of these molecules.

4.3. Using the Spin Hamiltonian

Abragam and Pryce [18] have developed a perturbation procedure for the calculation of splittings of a paramagnetic ion, which has found extensive application in electron spin resonance studies [19]. This method, which employs the so-called spin Hamiltonian, has been used for interpreting Mössbauer spectra of paramagnetic complex ions [20,21]. We will not give a complete discussion of the spin Hamiltonian, for which the reader is referred to specialized references, but rather outline the use of this method in Mössbauer spectroscopy.

A general spin Hamiltonian is the sum of energy operators

$$\begin{aligned}
H = &\ \beta(g_z H_z S_z + g_x H_x S_x + g_y H_y S_y) \\
&+ D\{S_z^2 - 1/3S(S + 1)\} + E(S_x^2 - S_y^2) \\
&+ A_z S_z I_z + A_x S_x I_x + A_y S_y I_y \\
&+ P\{I_z^2 - 1/3I(I + 1)\} + P'(I_x^2 - I_y^2) \\
&+ g_n \beta_n H_e I
\end{aligned} \tag{3}$$

where H_e is the external magnetic field and S is the "effective spin" of the paramagnetic ion, such that $2S+1$ is the multiplicity of the lowest group of electronic states. In simple cases S is equal to the ionic spin. The first line represents the interaction of the effective spin with the external field, whereas the last line represents the interaction of this field with the nuclear spin. β is the Bohr magneton and β_N the nuclear magneton. The second line expresses the coupling of electron orbitals to the lattice. D and E are related to the electrostatic ligand field. The third line expresses the coupling between the effective spin of the electrons and the nuclear spin. The fourth line expresses the quadrupole coupling of the nucleus.

Let us see now how the spin Hamiltonian is used for the study of an iron complex of biological importance.

The azide derivative of hemoglobin (Figure 16) contains a low-spin ferric ion ($S=1/2$) with a single hole in the lower orbital triplet. From electron spin resonance measurements [22] it was found that $g_x=1.70$, $g_y=2.20$, and $g_z=2.82$. The spin Hamiltonian for an effective spin $S=1/2$ reduces to the first line of Eq. (3) since the second term vanishes, and the remaining ones are dropped because the hyperfine structure is not resolved in most electron spin resonance spectra of iron compounds.

Using the spin Hamiltonian $H=\beta(g_z H_z S_z + g_x H_x S_x + g_y H_y S_y)$ with the ground-state wave function for the iron ion

$$\psi^+ = a|1a> + b|\zeta\beta> + c|-1a> \tag{4}$$
$$\psi^- = a|-1\beta> - b|\zeta a> + c|1\beta>$$

where $|1a>$, $|\zeta\beta>$, etc., are linear combinations of the d_{xy}, d_{xz}, and d_{yz} orbitals, Griffith [23] has established relations between the g values and the wave-function amplitudes a, b, and c. He fits the experimental g's with $a=0.841$, $b=0.099$, and $c=0.532$.

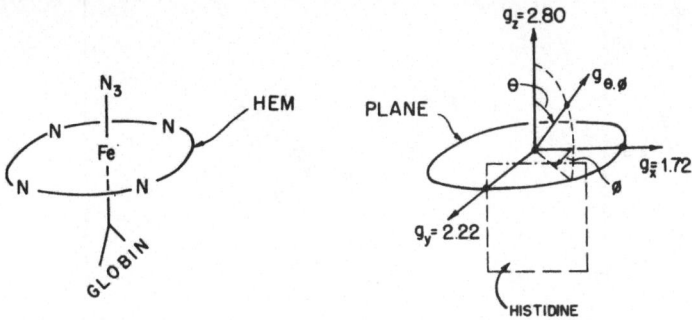

Figure 16. The structure and the three principal axes of g-value variation for myoglobin and hemoglobin azide.

The Mössbauer spectrum in the paramagnetic complex results from transition between excited and ground nuclear eigenstates determined by the nuclear spin Hamiltonians. In order to derive the form of these Hamiltonians from Eq. (3) we observe that the first and last term vanish since we have no external magnetic field. The second line vanishes for the effective spin $S=1/2$, and there remains the hyperfine and quadrupole coupling terms

$$H = A_z S_z I_z + A_x S_x I_x + A_y S_y I_y \qquad (5)$$
$$+ Q V_{zz}/4\{I_z^2 - 5/4 + \eta/3(I_x^2 - I_y^2)\}$$

In the nuclear ground state the quadrupole coupling vanishes, and Eq. (5) reduces to

$$H = A'_z S_z I_z + A'_x S_x I_x + A'_y S_y I_y \qquad (6)$$

where the prime is used to differentiate the ground from the excited nuclear state.

Using the wave functions [Eq. (4)] it is possible to express A, Q, and η as a function of a, b, and c [20]. Introducing numerical values of the parameters in Eqs. (5) and (6) and calculating the corresponding eigenstates, the line absorption spectrum shown in Figure 17 is obtained, which satisfactorily fits the observed Mössbauer spectrum [20].

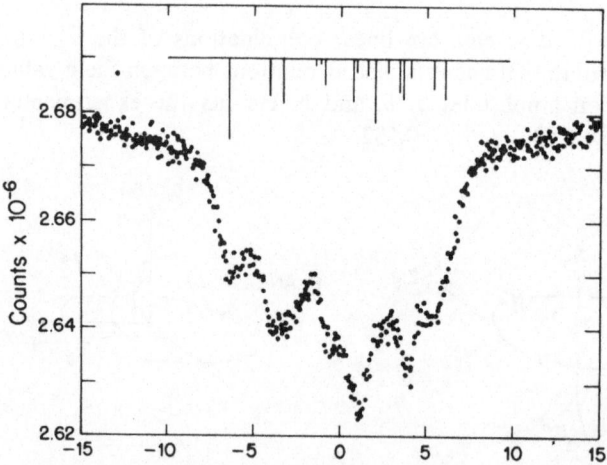

Figure 17. Comparison of predicted absorption lines and the 1.2°K hemoglobin azide data. The breadth of the observed absorption line is attributed to spin relaxation. Adapted from [20].

REFERENCES

1. V. I. Gol'danskii and R. H. Herber, Eds., *Chemical Applications of Mössbauer Spectroscopy* (Academic Press, New York, 1968).
2. J. Danon, *Lectures on the Mössbauer Effect* (Gordon and Breach, New York, 1968).
3. N. N. Greenwood, *Chemistry in Britain* 3, 56 (1967).
4. J. Danon, in *Physical Methods in Advanced Inorganic Chemistry*, H. A. O. Hill and P. Day, Eds. (Interscience Publishers, London, 1969).
5. F. Albert Cotton and G. Wilkinson, *Advanced Inorganic Chemistry* (Interscience, New York, 1966).
6. G. Kaindl, N. Potzel, F. Wagner, Ursel Zahn, and R. L. Mössbauer *Z. Physik.* 266, 103 (1969).
7. G. R. Hoy and F. de S. Barros, *Phys. Rev.* 139, A929 (1965).
8. J. Danon, *Rev. Mod. Phys.* 36, 454 (1964).
9. M. D. Lind, *J. Chem. Phys.* 46, 2010 (1967).
10. M. D. Lind, *J. Chem. Phys.* 47, 990 (1967).
11. R. R. Berrett and B. W. Fitzsimmons, *J. Chem. Soc.* A1967, 525.
12. D. B. Brown, D. F. Shriever, and L. H. Schwartz, *Inorg. Chem.* 7, 77 (1968).
13. E. Konig and K. Madeja, *Chem. Commun.* 1966, 61.
14. A. Ito, M. Suenaga, and K. Ono, *Tech. Rept. ISSP, Ser. A*, No. 289 (1967); *J. Chem. Phys.* 48, 3597 (1968).
15. R. L. Ingalls, *Phys. Rev.* 133, A787 (1964).
16. J. Danon and L. Iannarella, *J. Chem. Phys.* 47, 382 (1967).
17. P. T. Manahoran and H. B. Gray, *J. Am. Chem. Soc.* 87, 3340 (1965).
18. A. Abragam and M. H. L. Pryce, *Proc. Roy. Soc.* (*London*) A205, 135 (1951).
19. J. S. Griffith, *The Theory of Transition Metal Ions* (Cambridge University Press, 1961).
20. G. Lang and W. Marshall, *Proc. Phys. Soc.* 87, 3 (1966).
21. G. Lang and W. Marshall, *Mössbauer Effect Methodology*, Vol. 2 (Plenum Press, New York, 1966), p. 127.
22. J. F. Gibson and D. J. E. Ingram, *Nature* 180, 29 (1967).
23. J. S. Griffith, *Nature* 180, 31 (1957).

Chapter 7

Application to Organometallic Compounds

R. H. Herber

Rutgers University
New Brunswick, New Jersey

Mössbauer spectroscopy, used in conjunction with the other spectroscopic methods which are normally employed in structure elucidation, such as infrared, nuclear magnetic resonance (NMR), and x-ray diffraction techniques, provides a powerful tool for the study of organometallic compounds incorporating the nuclides which are suitable for such studies. Fortunately, the two elements which are most readily accessible for Mössbauer work by chemists—iron and tin—have a very extensive and varied organometallic chemistry and are representative in many ways of transition metals and nontransition metals, respectively. In the present chapter, some of the broad outlines of Mössbauer studies on organometallic compounds of iron and tin will be summarized, and some examples from the recent literature will be reviewed. However, this discussion is intended to be neither encyclopedic nor exhaustive, and the interested reader is referred to some of the reviews on this topic which have appeared in the literature [1,2].

In the present context, the qualification "organometallic" will refer to molecules having at least one metal-to-carbon bond, excluding, however, compounds which are usually classified as "inorganic," such as graphites, carbides, cyanides, and related materials. The metal-to-carbon bond may be either a *sigma bond* [e.g., as in $(CH_3)_4Sn$], a *pi bond* [e.g., as in $(\pi\text{-}C_5H_5)_2Fe$], or a related interaction in which the nearest neighbor atom to the metal involved is a carbon atom. As will be discussed in the following pages, such compounds have a set of related characteristics which make them appropriately a subject of integrated study. Broadly speaking, these characteristics include, *inter alia*: (a) relatively low Debye temperatures, (b) similar electron configurations and effective atomic numbers, (c) suitability for NMR and

Table 1. Partial Isomer-Shift Values

Ligand	Partial isomer shift[a]
$\diagdown C = O$	+0.084
$-C \equiv O$	+0.034
Fe — Fe	+0.008
S (one e^- donor)	+0.19
S (two e^- donor)	+0.017
NO	+0.105
Cl $\approx$ Br $\approx$ I	+0.34–0.35

[a] Relative to SNP at 296°K, mm/sec.

infrared study, (d) similar solubility properties, (e) similar synthetic origins, (f) similar thermal stabilities, etc.

1. ISOMER SHIFTS

1.1. Organoiron Compounds

The isomer shifts of such compounds generally fall into the range +0.1–1.0 mm/sec[1] and are thus similar to the values reported for other low-spin complexes of iron, such as $K_4Fe(CN)_6 \cdot 3H_2O(+0.213 \pm 0.005)$ [1], $K_3Fe(CN)_6$ (+0.136 $\pm$ 0.003) [1], and SNP itself (0 by definition).

Since virtually all metal organic compounds of iron have an effective atomic number (EAN) of 36, the Mössbauer isomer shifts—which reflect primarily the s-electron densities at the ^{57}Fe nucleus—can provide some detailed information concerning the bonding interaction between the metal atom and its associated ligands. Moreover, x-ray diffraction data have clearly established that for such compounds there is generally a characteristic bonding distance between the metal atom and a given ligand when this combination occurs in a variety of (related) compounds. Consequently, it may be expected that a given ligand will make a consistent (and possibly unique) contribution to the observed isomer shift in a Mössbauer spectrum. This idea was first systematically explored in an extensive study [3] of iron organic compounds which showed that a *partial isomer shift* could be ascribed to a particular ligand bonded to the iron atom. Typical values of some partial

---

[1] All iron isomer shifts discussed in the present chapter refer to shifts from the center of the sodium nitroprusside (SNP, Fe(CN)$_5$NO·2H$_2$O) spectrum at 296°K. This material, which is available as Reference Material 725 from the U.S. Bureau of Standards, has the following (approximate) shifts with respect to commonly used source matrices (mm/sec at room temperature): platinum +0.607, palladium +0.447, copper +0.479, chromium +0.075, iron +0.258, stainless steel +0.161–0.180.

isomer shifts are summarized in Table 1. For pi-bonding ligands, such as cyclopentadienyl groups, the partial isomer-shift concept needs to be modified to reflect the details of the bonding requirements of the other ligands. Specifically, although it might be inferred that the partial isomer shift of (π-C_5H_5) is just one half of the observed isomer shift for ferrocene, (π-C_5H_5)$_2$Fe, the value so calculated is not appropriate for compounds containing only a single cyclopentadienyl group. By reference to the NMR proton chemical shift τ, however, it is observed that a consistent set of partial isomer shifts for pi-cyclopentadienyl ligands can be calculated from the empirical relationship

$$\text{(partial isomer shift) } \pi\text{-}C_5H_5 = +0.223 \ (\tau\text{-}4.24) \text{ mm/sec}$$

There is an important consequence of the partial isomer-shift concept that should be borne in mind in connection with distinguishing between two chemically nonequivalent iron atoms in a single molecule. It is clear that such distinct iron atoms may, in fact, have the same isomer shift by an appropriate contribution from the various ligands and that the Mössbauer spectrum of such a compound may thus give evidence of what is (apparently) only one kind of iron atom. For example, the two iron atoms in (π-C_5H_5) Fe(CO)$_2$CH$_2$COFe(CO)$_2$(π-C_5H_5) both have essentially the same isomer shift and cannot be distinguished in the Mössbauer spectrum. It is also clear that two iron atoms which have the same *nearest neighbor* environment around the metal atom, but different structures, will have in fact the same isomer shift. This point has been confirmed for two pairs of cyclopentadienyl iron complexes which can be obtained in both *cis* and *trans* conformation [4], as well as in the case of two octahedral *cis, trans* compounds [5].

1.2. Organotin Compounds

The isomer shifts for most organotin compounds fall in the range of $+1.2$–1.8 mm/sec with respect to SnO_2, so that these compounds can be thought of as either derived from Sn^{4+} or from Sn, although the latter view is more consistent with the high covalency in tin–carbon bonds expected from electronegativity considersations.

The correlation between the ^{119}Sn isomer shifts of tin tetrahalides and the electronegativity of the halide, first noted by Gol'danskii *et al.*, [6] has prompted a similar study of organotin compounds, which showed [7], however, that the isomer-shift electronegativity correlation is strongly perturbed by the stereochemistry of the complex. A much better relationship between these two parameters is obtained for the structurally related phthalocyaninotin complexes of the type PcSnX$_2$(X=F, Cl, Br, I, or OH) [8], and such studies may shed considerable light on the appropriate group electronegativity of polyatomic ligands (see however [45, 46]).

With respect to the isomer shifts of a series of homologous alkyl tin compounds, it has been noted [9] that molecules having at least one $Sn-CH_3$ bond consistently show isomer shifts which are ~ 0.18 mm/sec more negative than related compounds in which the alkyl groups are C_2 or longer. Although it is tempting to ascribe these differences to the respective electronegativity of hydrogen and the methyl (or methylene) group, more detailed study is needed to account for the observed effect in a quantitative manner.

2. QUADRUPOLE SPLITTING

2.1. Organoiron Compounds

As already noted, organoiron compounds are preponderantly diamagnetic, with an effective atomic number of 36. This electron configuration can be thought of as arising from a bonding combination of elemental iron $(Z=26)$ and 10 electrons derived from (neutral, molecular) ligands. In contrast to low-spin octahedral ferric $(EAN=23+12=35)$ or low-spin octahedral ferrous $(EAN=24+12=36)$ complexes in which the ligands can assume cubic symmetry about the metal atom, it is not possible to dispose the ligands in organoiron compounds around the metal atom in such a way as to obtain a cubic charge distribution. It is for this reason that essentially all organoiron compounds show a nonzero quadrupole splitting, although the magnitude of such splitting may not be large enough to permit a clear resolution of the Mössbauer effect doublet. It should be borne in mind, however, that while this symmetry argument pertains to the *minimal* quadrupole splitting associated with such compounds, the major contribution to the quadrupole splitting lies in the nonequivalent population of the $3d$ orbitals.[2]

A dramatic illustration of this point may be found in the Mössbauer data for ferrocene and ferricinium salts, which have been examined in detail in the elegant experiments of Collins [11] in conjunction with a determination of the sign of the EFG tensor in these compounds using the magnetic field technique of Ruby and Flinn [12]. The large quadrupole splitting observed in $(\pi-C_5H_5)_2Fe$ (2.37 mm/sec at 78°K, 2.34 mm/sec 295°K [13]) due to the $3d_0$ orbital population, collapses almost completely with the removal of one electron from the metal atom.

This characteristic behavior has recently been exploited in the elucidation of the bonding in two iron carbollide complexes which were first isolated by Hawthorne and his coworkers [14]. The crystal structure data

---

[2] As has recently been pointed out by Harris [10], the field at the metal atom nucleus due to a single $3d$ electron is at least three times that due to a $4p$ electron, and in the iron region of the first transition series, this ratio may be as large as five or six.

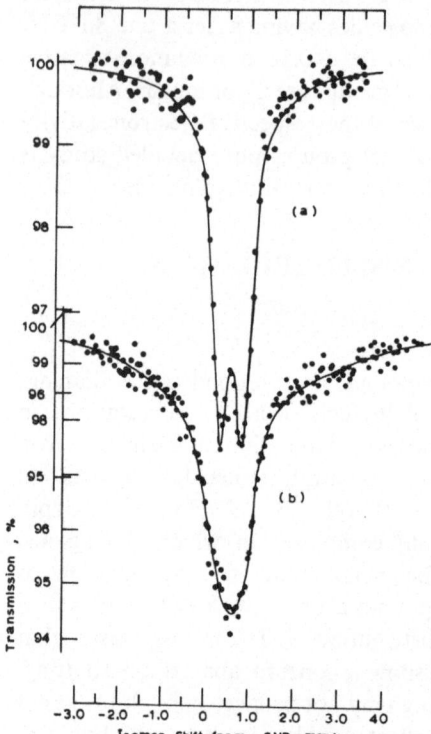

Figure 1. Mössbauer spectra for (a) $(\pi\text{-}C_5H_5)Fe(C_2B_9H_{11})$ and (b) $(CH_3)_4NFe$ $(C_2B_9H_{11})_2$, both at 140°K. The source used in obtaining these data was Pd(^{57}Co) at 296°K. The isomer-shift scale is with respect to the centroid of sodium nitroprusside at 296°K.

[15] for the compound $\pi\text{-}C_5H_5FeC_2B_9H_{11}$ shows that the iron atom is "sandwiched" between the pi-cyclopentadienyl group and a five-membered ring fragment of the carbollyl icosahedron which contains two carbon atoms and three boron atoms. The structure of the bis-carbollyl complex is such that the metal atom is sandwiched between two C_2B_3 moieties in an analogous manner.

The Mössbauer results on these two complexes [16], together with data for related cyclopentadienyl complexes, is summarized in Table 2 and shown graphically for III (a) and IV (b) in Figure 1. There are several conclusions which can be inferred from these data: (a) the isomer shifts of II, III, and IV show that the formal oxidation state of the metal atom in these complexes is 3+ and that the carbollyl complexes are nominally analogous to ferricinium complexes; (b) the same conclusion arises from the observation of the small quadrupole splitting in IV [the quadrupole splitting in III could not be resolved under the experimental conditions employed, however, the line width Γ is ~ 1.33 mm/sec at 140°K and may thus involve a hyperfine splitting on the same order as that observed in typical ferricinium salts]; (c)

Table 2. Mössbauer Parameters for $Fe(C_2B_9H_{11})^-$ and Related
Compounds[a]

Absorber		Temperature, °K	δ, mm/sec[b]	ΔE_Q, mm/sec
$(\pi\text{-}C_5H_5)_2Fe$	(I)	77	$+0.773\pm0.020$	2.417 ± 0.020
$[(\pi\text{-}C_5H_5)_2Fe^+]Br^-$	(II)	77	$+0.655\pm0.040$	[c]
$[(CH_3)_4N][Fe(C_2B_9H_{11})_2]$	(III)	140	$+0.630\pm0.030$	—
$(\pi\text{-}C_5H_5)Fe(C_2B_9H_{11})$	(IV)	140	$+0.608\pm0.030$	0.529 ± 0.030
$[(\pi\text{-}C_5H_5)Fe(C_5H_4)]_2CH^+BF_4^-$	(V)	100	$+0.785\pm0.020$	2.111 ± 0.020
$[(\pi\text{-}C_5H_5)Fe(C_5H_4)]_3C^+ClO_4^-$		80	$+0.75^d$	2.05^d
$[\pi\text{-}C_5H_5)Fe(C_5H_4)]_2CH^+ClO_4^-$		80	$+0.67^d$	2.05^d

a The data for this table are taken largely from [16], except as noted.
b With respect to SNP at 296°K.
c Not resolved. The line width is 0.556 ± 0.040 mm/sec.
d Data taken from [2].

the bonding characteristics of the C_2B_3 fragment of the carbollide icosahedron are essentially the same as those of the $\pi\text{-}C_5H_5$ ring, implying a
delocalization of the electrons over the whole ring as in the ferrocene analogs.
The detailed analysis of the nqr data for these and related complexes has
been given by Harris [10] and the conclusions obtained therefrom are in
good agreement (and are complementary) with those extracted from the
Mössbauer experiments.

2.2. Organotin Compounds:

The systematics of the quadrupole hyperfine interaction for the ^{119}Sn
Mössbauer resonance in organometallic compounds show a markedly

Table 3. Tetrahedral Organotin Compounds in Which $\Delta E_Q = 0^a$

Absorber	Temperature, °K	δ, mm/sec	ΔE_Q, mm/sec
$(CH_3)_4Sn$	78	$+1.30$	0
$(C_2H_5)_4Sn$	78	$+1.30$	0
$(C_3H_7)_4Sn$	78	$+1.30\text{-}1.33$	0
$(C_4H_9)_4Sn$	78	$+1.30\text{-}1.35$	0
$(C_6H_{11})_4Sn$	78	$+1.52$	0
$(C_6H_5)_4Sn$	77–80	$+1.10\text{-}1.35$	0
SnH_4	78	$+1.27$	0

a The data for this table are taken from [2]. The ^{119}Sn isomer shifts here—and else-
where in the present chapter—have been recalculated with respect to the center of
the SnO_2 absorption peak at room temperature. Recent precision measurements in
this laboratory have shown that $\delta (SnO_2) = \delta (BaSnO_3)$ within the experimental
error of ±0.035 mm/sec. Characteristic shifts with respect to some other reference
absorbers, which have been quoted in the literature, are: α-Sn (grey, tetrahedral):
$+1.55$ mm/sec; Mg_2Sn: $+1.80$ mm/sec; Pd (dilute alloy, $\sim 3\%$ Sn): $+1.46$ mm/sec.

Table 4. Compounds which Lack Cubic Symmetry but in which
Quadrupole Splitting Is Not Observed

Absorber	Temperature, °K	δ mm/sec[a]	ΔE_Q, mm/sec
$(CH_3)Sn(C_6H_5)_3$	78	+1.19	0
$(CH_3)_3SnC_6H_5$	78	+1.08	0
$(C_2H_5)_3SnCH_3$	78	+1.35	0
$(CH_3)SnH_3$	78	+1.24	0
$(CH_3)_3SnH$	78	+1.24	0
$(C_6H_5)_3SnH$	78	+1.39	0
$(C_2H_5)_3Sn-Sn(C_2H_5)_3$	78	+1.45–1.55	0
$(C_2H_5)_3Sn\ Sn(C_6H_5)_3$	78	+1.45–1.55	0
$(C_6H_5)_3SnLi$	78	+1.40	0
$[(C_6H_5)_3Sn]_2Sn$	80	+1.33	0
$[(C_6H_5)_3Sn]_4Ge$	80	+1.13	0
$[(C_4H_9)_2Sn]_n$	78	+1.55	0

[a] Relative to SnO_2.

different behavior from that observed for organoiron molecules. The most common stereochemical configuration of compounds of the type R_4Sn, is tetrahedral, making use of four equivalent $5s5p^3$ hybrid orbitals of the metal atom. As expected, the quadrupole splitting in such compounds is zero since the bonding orbitals have cubic symmetry with respect to the metal atom lattice site, and hence the EFG tensor vanishes at the tin nucleus. The Mössbauer parameters of typical compounds of this type are summarized in Table 3.

A more extensive study of organotin compounds [17] has shown that a large number of such molecules, for which *a priori* expectation is for a nonzero quadrupole hyperfine interaction to be present, in fact show no splitting of the Mössbauer resonance within the experimental error limits usually associated with such measurements.[3] Among such compounds are binuclear alkyl and aryl compounds having Sn–Sn bonds, alkyl and aryl stannanes, molecules in which a triphenyl-tin moiety is bonded to a nontransition metal, and a number of organotin polymers. Several examples of molecules in which a nonvanishing field gradient is expected, but in which no quadrupole splitting is experimentally observed, are summarized in Table 4.

It was first pointed out by Gibb and Greenwood [18] that in molecules

---

[3] It should be noted in passing that a consideration of line shapes in deciding whether or not a splitting of the Mössbauer resonance in fact occurs, can be very misleading. The envelope of the sum of two Lorentzian functions, $I(v) = [I(v_0)]/[I+(4(v-v_0)/\Gamma_a)]^2$, when the two lines are separated by $\leq 1/2(2\Gamma_{nat})$ is so nearly Lorentzian that it is difficult to distinguish the nearly complete overlap of two absorption lines from the presence of a single unsplit absorption.

Table 5.   Effect of Lone Pair Electrons on the Magnitude of the
Quadrupole Interaction in Organotin Compounds

Absorber	Temperature, °K	δ mm/sec[a]	ΔE_Q, mm/sec
$(CH_3)_4Sn$	78	+1.30	0
$(CH_3)_3SnF$	78	+1.18–1.26	3.47–3.77
$(CH_3)_3SnCl$	78	+1.40–1.60	3.09–3.67
$(CH_3)_3SnH$	78	+1.24	0
$(CH_3)_3SnOH$	78	+1.07	2.71
$(CH_3)_3Sn-N=N\equiv N$	78	+1.34	3.23
$(CH_3)_3SnC_6H_5$	78	+1.08	0

[a] Relative to SnO_2.

in which no quadrupole splitting is observed, the nearest neighbor ligand atoms lack lone electron pairs and that, on the other hand, molecules in which one, two, or three of the atoms bonded to the tin atom have such lone pairs, a large quadrupole interaction would be observed in the Mössbauer spectrum. This effect is clearly evident in the series of compounds summarized in Table 5.

The lone pair rule can be understood [19] on the basis of the imbalance in the $5p$ (and $6p$) orbital occupation and has been generalized as follows: (a) all molecules with a large field gradient show a large quadrupole splitting; (b) molecules in which the three p orbitals remain triply degenerate show zero quadrupole splitting; (c) molecules in which the degeneracy of the three p orbitals is removed by a p_π or pseudo p_π interaction (or by steric requirements such as in trigonal bipyramid symmetry) will show large quadrupole splittings. Further refinement of the molecular orbital calculations reported by Greenwood, Perkins, and Wall [19] will be required to elucidate the significance of the minimum field gradient which can be observed.

Moreover, ligands with lone pairs such as the halogens, nitrogen, oxygen, etc., tend to stabilize the tin atom in a trigonal bipyramid configuration, especially for molecules of the general formula R_3SnX. Thus, in trialkyl tin compounds of the type $(CH_3)_3SnF$, $(CH_3)_3SnCN$, and $(CH_3)_3Sn-Clpy$, the R_3Sn moiety tends to assume a planar (or nearly planar) configuration. Using Bent's rule [20], the s character in the metal ligand bonds will be concentrated in the bond to the more electropositive ligand, i.e., the R group, while the p character will be concentrated in the Sn–X bond, i.e., F, CN, py, etc. The approximate 180° bond angle in the X...Sn...X bond, i.e., the C_{3v} symmetry about the Sn atom, gives rise to an imbalance in the p orbital population and hence to large field gradients.

There are only a few examples of organotin compounds in which the lone pair rule appears at first glance to be violated. The compound $(C_6H_5)_3$ $Sn(C_6F_5)$ is reported [21] to have a ΔE_Q of 1.1 mm/sec at 80°K. It is likely that in this compound the electric field gradient at the tin atom nucleus arises from an overlap between the lone pairs on the *ortho* fluorine atoms of the perfluorophenyl group and the $6p$ orbitals of the metal, since this appears to be sterically allowed. Gol'danskii and his coworkers [22] have reported extensive data on barenyl derivatives of tin in which the moiety —$C(B_{10}H_{10})CR$ ($R=H,C_3H_7,C_6H_5$, etc.) is bonded to an alkyl or aryl tin fragment. Such compounds show quadrupole splittings in the range 0.95– 1.70 mm/sec depending on the nature of R. The two carbon atoms, which are part of a carborane icosahedron, are directly involved in the bonding to the metal atom, and the whole ligand is thought to be strongly electron withdrawing [23], thus giving rise to the quadrupole interactions which are observed.

Finally, it is worth noting that both alkyl tin cyanides [24,25] and thiocyanates [25] show large quadrupole splittings at liquid-nitrogen temperature (Table 6), although the infrared evidence suggests a normal carbon– nitrogen triple bond in the cyanide group. In these molecules, even though there are (formally) no lone pairs on the nearest neighbor atom, the $6p$ orbitals of the tin presumably overlap sufficiently with the π-electron density in the ligand framework to give rise to the hyperfine interactions which are observed.

Although most organotin compounds show a coordination of four or five around the tin atom, there are a number of complexes which have been characterized in which the metal atom has a coordination number of six and in which octahedral or quasioctahedral structures are achieved. The generali-

Table 6. Mössbauer Parameters for Organotin Cyanides and Thiocyanates[a]

Absorber	Temperature, °K	δ, mm/sec	ΔE_Q, mm/sec
$(CH_3)_3SnCN$	96	+1.39	3.12
$(C_2H_5)_3SnCN$	96	+1.41	3.19
$(C_4H_4)_3SnCN$	96	+1.37	3.27
$(CH_3)_3SnSCN$	96	+1.40	3.77
$(C_2H_5)_3SnSCN$	96	+1.57	3.80
$(C_4H_9)_3SnSCN$	96	+1.60	3.69
$(C_6H_5)_3SnSCN$	96	+1.35	3.50
$(CH_3)_2Sn(SCN)_2$	96	+1.48	3.87
$(C_2H_5)_2Sn(SCN)_2$	96	+1.56	3.96
$(C_4H_9)_2Sn(SCN)_2$	96	+1.56	3.88
$(C_5H_9)Sn(SCN)_3$	96	+1.43	1.46

[a] Data from [25]. δ relative to SnO_2.

zations concerning quadrupole splitting in such molecules are complementary to those referred to above and have their origins in the same considerations of p orbital populations as have already been discussed. Six coordinate tin compounds in which at least one ligand *lacks* a lone pair, will show a large quadrupole splitting. For example, the splitting in $(C_4H_9)_2$ Sn $(OCOCH_3)_2$ is 3.5 mm/sec and in $(CH_3)_2$ Sn $(OCHO)_2$ it is 4.72 mm/sec at 78°K. Both of these compounds are assumed to contain a six coordinate tin atom with the formate or acetate ligand acting as a bidentate moiety to the same or a neighboring metal atom.

On the other hand, if all six of the nearest neighbor atoms have lone pairs, then quadrupole splitting will be unresolvable in the resultant spectra, even if the six nearest neighbor atoms are nonequivalent [27,45]. The absence of a resolvable quadrupole splitting is expected in complexes having O_h symmetry, such as $SnCl_6^{2-}$, $SnBr_6^{2-}$, etc. However, even when two of these halogen ligands are replaced by other atoms so that the resulting species clearly lacks octahedral symmetry about the metal atom, $\Delta E_Q = 0$ except when one or more of the ligands is an alkyl (or aryl) group. This generalized observation can clearly be used in the formulation of model structures for organotin compounds and specifically to elucidate the nature of the bonding in molecules which associate to linear or three-dimensional polymers in solution.

3. CONFORMATIONAL STUDIES

Mössbauer spectroscopy can make a unique contribution to the study of the possible conformational changes of organometallic compounds when these are subjected to changes in environment. The need for such studies arises from the fact that the most unambiguous structural information is derived from x-ray diffraction data on single crystal samples, while a great deal of structural data on such molecules is inferred from molecular spectroscopic measurements on solutions, i.e., NMR and infrared measurements, primarily. Since Mössbauer measurements can be carried out both on neat solids and (frozen) solutions, this method is able to bridge the gap between diffraction and spectroscopic measurements and—in some cases—to resolve the apparent differences in the structural information so derived. In the present discussion, two instances in which Mössbauer spectroscopy has aided in the study of the structural integrity of organometallic compounds will be briefly reviewed.

3.1. Cyclooctatetraene Iron Tricarbonyl [COTFe(CO)₃]

This compound was first reported in 1959 [28–30]. The NMR spectrum of the molecule consists of a single sharp proton resonance at room temperature, and it was thus concluded [31] that the complex is an "open-faced sandwich" with the cyclooctatetraene ring approximately planar and symmetrically involved in the bonding to the $Fe(CO)_3$ moiety, thereby making all eight protons equivalent. An x-ray crystallographic study [32] showed, however, that in the solid state, the $Fe(CO)_3$ moiety is asymmetrically situated above a C_4H_4 fragment of the ring, and that the remaining hydrocarbon framework is separated by distances which preclude a direct bonding interaction.

On this basis, the room temperature NMR spectrum was reinterpreted as implying the operation of an averaging mechanism which makes all of the eight protons equivalent on a time scale short compared to that of an NMR scan, i.e., $\sim 10^{-6}$ sec. The expectation that this averaging process may be slowed down by reducing the temperature of the sample during the spectroscopic examination was confirmed, and the low-temperature NMR spectra [33–35] show a splitting of the proton resonance into two structured resonance peaks symmetrically displaced from the room-temperature resonance. The low-temperature NMR spectrum was interpreted differently by different groups, and the various suggestions offered can be conveniently characterized as having the $Fe(CO)_3$ fragment bonded to: (a) a biplanar COT ring acting as a 1,3-diene [3], (b) a tub-shaped COT ring acting as a 1.3-diene [35], and (c) a tub-shaped COT ring acting as a 1,5-diene [34]. The apparent disagreement concerning the structure of the low-temperature-solution conformation of COT $Fe(CO)_3$ was subsequently resolved by a series of Mössbauer measurements carried out on neat solids and on frozen solution samples of COT $Fe(CO)_3$ and related molecules [36].

The solvents chosen for such studies must meet several requirements. In addition to chemical inertness and good solvating power for the solutes in question, they should be glass-formers at low temperature (so that the solvent accommodates the structure of the solute rather than imposing a crystalline environment on the latter), they should have a reasonably high lattice temperature (so that the resonance effects will be as large as possible at the temperature of the measurement, usually liquid-nitrogen temperature), and they should be readily available in high purity. With respect to the COT $Fe(CO)_3$ problem, these criteria are met by EPA and methyl hydrofuran (MTF), and the Mössbauer parameters of several iron tricarbonyl complexes [$RFe(CO)_3$] in these and related solvents are summarized in Table 7. These data are also represented graphically in Figure 2, which is a correlation diagram (δ vs ΔE_Q) for the iron tricarbonyl compounds as neat solids and as frozen solution samples. From this figure, it is evident that the quadrupole splitting parameter (which is most sensitive to variations in the structures that are being considered) of COT $Fe(CO)_3$ is essentially the same in the

Table 7. Mössbauer Parameters for RFe(CO)₃ Compounds

Radical	Solvent[a]	δ, mm/sec[b]	ΔE_Q, mm/sec
Cyclooctatetraene (I)	....	+0.31	1.24
	EPA	+0.34	1.16
	$C_6H_5NO_2$	+0.36	1.21
	n-C_8H_{18}	+0.36	1.23
	MTF	+0.35	1.20
1,8-Dicarbomethoxy-COT, $C_8H_6(CO_2CH_3)_2$ (III)	....	+0.33	1.27
COT lactone (IV)	....	+0.33	1.25
	MTF	+0.33	1.17
Norbornadiene (II)	....	+0.29	2.15
Cycloocta-1,5-diene (V)	....	+0.23	1.83
$C_8H_8[Fe(CO)_3]_2$	....	+0.24	1.32
Cyclobutadiene	....	+0.28	1.54
	EPA	+0.26	1.55

[a] EPA = 16/42/42 % v/v, ethanol, 1-propanol, diethyl ether; MTF = methyltetrahydrofuran.
[b] Relative to SNP at 294°K.

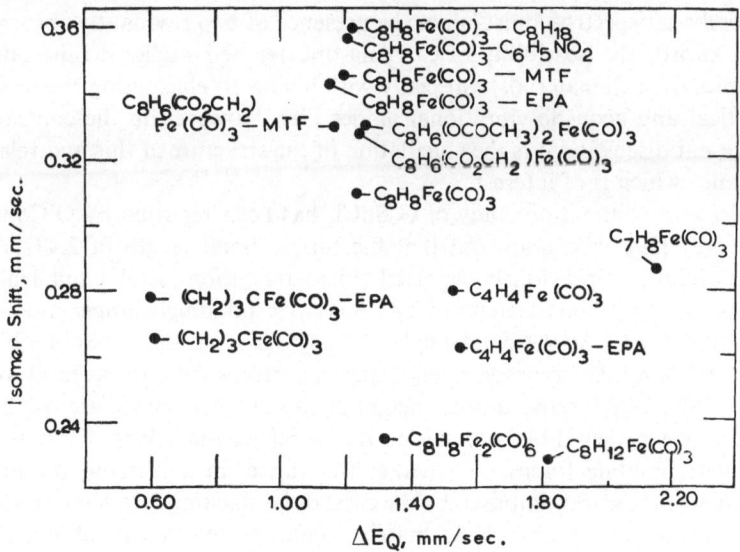

Figure 2. Correlation diagram showing Mössbauer parameters for a variety of organometallic Fe(CO)₃ complexes in which the metal is bonded to a C_4H_4 fragment.

solid and in EPA, MTF, nitrobenzene, or n-octane solution, and identical—
within experimental error—to the quadrupole splitting in the binuclear
complex COT[Fe(CO)$_3$]$_2$. Moreover, this parameter in the COT complexes
is markedly different from the value observed for C_8H_{12} Fe(CO)$_3$, a compound which is known from other evidence to be a 1,5-diene complex of iron
tricarbonyl, as well as being very much smaller than the value of the ΔE_Q
parameter observed for norbornadienyl iron tricarbonyl [C$_7$H$_8$Fe(CO)$_3$] in
which the pseudooctahedral structure of Dickens and Lipscomb [32] cannot
be realized.

On the basis of these and related NMR and infrared measurements it is
concluded that the structural assignment of Winstein et al. [33] is the most
tenable one and that the configuration of COT Fe(CO)$_3$ in solution is
essentially the same as that observed more directly by x-ray methods for the
solid.

3.2. [π-C$_5$H$_5$ Fe(CO)$_2$]$_2$ SnCl$_2$ and Related Molecules

In 1964, Bonati and Wilkinson reported [31] on the reaction between
the dimer [π-C$_5$H$_5$ Fe(CO)$_2$]$_2$ (G$_2$) and a variety of stannous salts and organotin compounds. Among the products which they isolated is [π-C$_5$H$_5$ Fe(CO)$_2$]$_2$
SnCl$_2$ (G$_2$SnCl$_2$) for which an early structural assignment, based largely on
infrared evidence, could be made. This compound is of considerable interest
to Mössbauer spectroscopists since the presence of two resonantly absorbing
nuclei affords the possibility of carrying out detailed studies on the lattice
dynamics of such materials, especially with a view to elucidating the mixing
of optical and acoustic vibrational modes [38]. However, in the context of
the present discussion it is the elucidation of the structure of this and related
molecules which is of interest.

An x-ray diffraction study of G$_2$SnCl$_2$ has been reported by O'Connor
and Corey [39], who point out that the Sn–Fe bond length of 2.492 Å is
shorter than any previously reported tin to transition-metal bond length,
and that the Sn–Cl bond length of 2.43 Å is correspondingly longer. It is also
noted that the Cl–Sn–Cl bond angle of $94.1\pm0.6°$ and the Fe–Sn–Fe bond
angle of $128.6\pm0.3°$ represent very large departures from tetrahedral symmetry (109° 24′) for the nearest neighbor atoms. An immediate question
thus arises whether these anomalous values reflect the effects of the intra-
molecular bonding forces, or whether they arise, in fact, from the inter-
molecular interactions imposed by the crystalline stacking forces in the solid.
By exploiting the frozen glassy matrix technique referred to above, it is
obvious that detailed Mössbauer effect measurements can be expected to
elucidate this point.

Such measurements on G$_2$SnCl$_2$, GSnCl$_3$, and related organometallic

Table 8. Summary of Mössbauer Data for [π-C₅H₅Fe(CO)₂]₂SnCl₂ and Related Molecules

Absorber	Matrix[a]	Resonance	δ, mm/sec[b]	ΔE_Q, mm/sec
[π-C₅H₅)Fe(CO)₂]₂	Neat	Fe	+0.47±0.01	1.91±0.01
[(π-C₅H₅)Fe(CO)₂]₂SnCl₂	Neat	Fe	+0.36±0.01	1.68±0.01
	MTHF	Fe	+0.36±0.01	1.68±0.01
	p	Fe	+0.35±0.01	1.74±0.05
	Neat	Sn	+1.95±0.02	2.38±0.02
	p	Sn	+1.96±0.02	2.25±0.02
			+1.31±0.02	2.29±0.02
(π-C₅H₅)Fe(CO)₂SnCl₃	Neat	Fe	+0.40±0.01	1.86±0.01
	MTHF	Fe	+0.39±0.01	1.76±0.01
	p	Fe	+0.38±0.01	1.78±0.01
	Neat	Sn	+1.74±0.02	1.77±0.02
	MTHF	Sn	+1.54±0.05	1.64±0.05
	p	Sn	+1.66±0.02	1.78±0.02
(π-C₅H₅)Fe(CO)₂Sn(C₆H₅)₃	Neat	Fe	+0.37±0.01	1.83±0.01
	p	Fe	+0.35±0.01	1.79±0.01
	Neat	Sn	+1.50±0.02	0
	p	Sn	+1.50±0.02	0
[(π-C₅H₅)Fe(CO)₂]₂GeCl₂	Neat	Fe	+0.36±0.01	1.66±0.01

[a] MTHF, 2-methylfuran; p, poly(methyl methacrylate).
[b] For Fe, relative to SNP; for Sn, relative to SnO₂.

compounds have been carried out [40] and are summarized in Table 8.[4] From the data for the ^{57}Fe and ^{119}Sn resonances in GSnCl₃ and G₂GeCl₂ it is clear that the isomer shifts and quadrupole splittings are essentially invariant with respect to going from the neat solid to frozen solution samples. From this it is clear that the anomalous bond angles discussed by O'Connor and Corey for G₂SnCl₂ and by Bush and Woodward [41] for G₂GeCl₂ arise from metal ligand interactions themselves, and are not the consequence of the stacking of these molecules into a crystalline solid.

The ^{119}Sn resonance spectrum of G₂SnCl₂ shows a markedly different behavior, as shown in Figure 3. The sharp doublet observed for the neat solid becomes a broad absorption spectrum which can be resolved into two quadrupole doublets of essentially identical splitting but rather different isomer shifts. The two doublets are indicative of the presence of two different tin atoms which are associated with two rotational conformers of G₂SnCl₂ in solution. The existence of such conformers has been previously

---

[4] The use of polymethylmethacrylate (p in the Matrix column of Table 8) has been extensively investigated by Y. Goscinny and S. Chandra in these laboratories. The major advantage of this matrix is its high lattice temperature, which permits the precision determination of weak resonances even at liquid-nitrogen temperature.

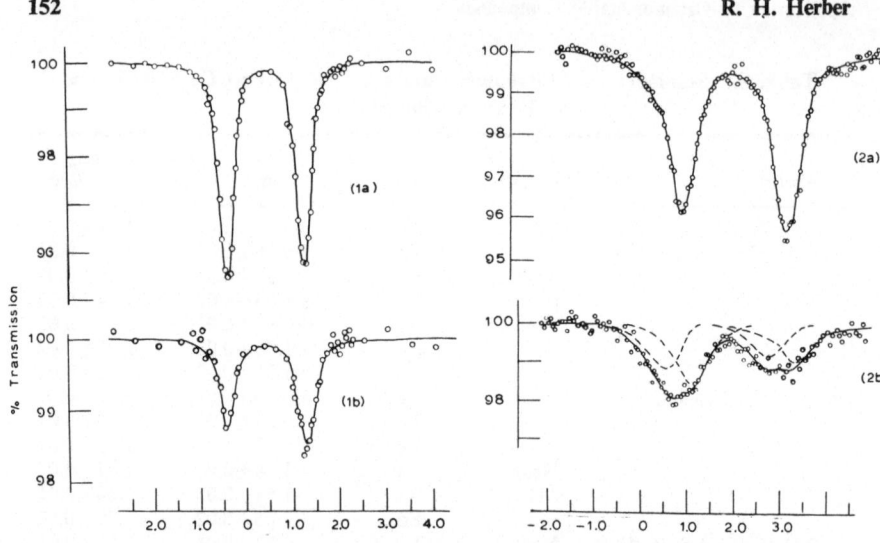

Figure 3. Mössbauer spectra for $[(\pi\text{-}C_5H_5)Fe(CO)_2]_2SnCl_2$. The upper two curves are ^{57}Fe resonances: (la) is the neat-solid data; (1b) is the polymethylmethacrylate matrix-sample data. The lower two curves are the ^{119}Sn resonances: (2a) is the neat-solid data; (2b) is the polymethylmetahcrylate matrix-sample data.

inferred from infrared evidence [42–44], and the present results provide an independent confirmation of the earlier speculation.

4. CONCLUSION

In the foregoing discussion an attempt has been made to outline some of the general aspects of Mössbauer spectroscopy as it can be applied to the study of organometallic compounds. Information concerning the bonding, symmetry, and architecture of such molecules can be derived by the methods outlined in the earlier discussions in this volume and in the literature references cited in the present chapter. Finally, a brief discussion of two applications of Mössbauer spectroscopy to the study of the integrity of molecular conformation has been presented, and the obvious role of this technique as a bridge between x-ray diffraction methods on single crystals and structural information obtained on solution samples by infrared and NMR techniques has been indicated.

There are a number of other measurements of Mössbauer spectroscopic parameters which can elucidate the nature of the structure and bonding in organometallic compounds, which—because of obvious limitations—have not been discussed in the present chapter. Among these are a variety of temperature dependence measurements, such as those of the recoil-free fraction, the isomer shift, quadrupole splitting, etc.; magnetic hyperfine interactions below the paramagnetic transition temperature; effect o concentration (in frozen solution measurements) on the recoil-free fraction; association–dissociation equilibria in solvents of varying polarity; Möss-bauer double resonance experiments; etc. Such measurements can serve to clarify the details of the chemical behavior and molecular architecture of organometallic molecules, and thus may be expected to be exploited in the future in continuing studies in this field.

Finally, it is worthwhile reemphasizing the point made at the beginning of this chapter which attempts to place the technique of Mössbauer spectro-scopy into its proper (and appropriate) context as a method for studying the chemical properties of organometallic molecules. Used in conjunction with x-ray diffraction, NMR, infrared spectroscopy, and the many other methods used in this field, it can make a significant and unique contribution to our understanding of this class of compounds. The present discussion has been intended to indicate, albeit briefly and superficially, some of the results which have already been obtained.

ACKNOWLEDGMENTS

The author is indebted to his students and associates for many stimulating discussions and ideas which have made such a major contribution to the development of this field, and without whom very little progress could have been made. This work has been supported in part by the U. S. Atomic Energy Commission (Document NYO-2472-56), the Petroleum Research Fund adminstered by the American Chemical Society, and the Research Council at Rutgers University, The State University of New Jersey. This support is herewith gratefully acknowledged.

REFERENCES

1. E. Fluck, in *Chemical Applications of Mössbauer Spectroscopy*, V. I. Gol'danskii and R. H. Herber, Eds. (Academic Press, New York, 1968), Chap. 4.
2. V. I. Gol'danskii, V. V. Khrapov, O. Yu. Okhlobystin, and V. Ya. Rochev, in *Chemical Applications of Mössbauer Spectroscopy*, (Academic Press, New York, 1968), Chap. 6. See also J. J. Zuckerman, *Adv. Organomet. Chem.* Vol 9, (Academic Press, New York, 1970).
3. R. H. Herber, R. B. King, and G. K. Wertheim, *Inorg. Chem.* 3, 101 (1964).
4. R. H. Herber and R. G. Hayter, *J. Am. Chem. Soc.* 86, 301 (1964).
5. R. R. Berrett and B. W. Fitzsimmons, *Chem. Commun.* 1966, 91.
6. V. I. Gol'danskii *et al.*, *Dokl. Akad. Nauk. SSSR* 147, 127 (1962).
7. R. H. Herber, H. A. Stöckler, and W. T. Reichle, *J. Chem. Phys.* 42, 2447 (1965).
8. H. A. Stöckler, H. Sano, and R. H. Herber, *J. Chem. Phys.* 45, 1182 (1966); 46, 2020 (1967).
9. R. H. Herber and G. I. Parisi, *Inorg. Chem.* 5, 769 (1966); R. H. Herber and Y. Goscinny (unpublished observations).
10. C. B. Harris, *Inorg. Chem.* 7, 1517 (1968).

11. R. L. Collins, *J. Chem. Phys.* **42**, 1072 (1965).
12. S. L. Ruby and P. A. Flinn, *Rev. Mod. Phys.* **36**, 351 (1964).
13. G. K. Wertheim and R. H. Herber, *J. Chem. Phys.* **38**, 2106 (1963); U. Zahn, P. Kienle, and H. Eicher, *Proceedings of the Second International Conference on the Mössbauer Effect*, D. M. J. Compton and A. H. Schoen, Eds. (John Wiley and Sons, New York, 1962), p. 271; *Z. Physik.* **166**, 220 (1962).
14. M. F. Hawthorne, D. C. Young, and P. A. Wegner, *J. Am. Chem. Soc.* **87**, 1818 (1965); M. F. Hawthorne and T. D. Andrews, *ibid.*, **87**, 2496 (1965); M. F. Hawthorne and R. L. Pilling, *ibid.*, **87**, 3987 (1965).
15. A. Zalkin, D. H. Templeton, and T. E. Hopkins, *J. Am. Chem. Soc.* **87**, 3988 (1965).
16. R. H. Herber, *Inorg. Chem.* **8**, 174 (1969).
17. R. H. Herber, *Intern. At. Energy Agency, Tech. Rept. Ser.* **50**, 121 (1966).
18. T. C. Gibb and N. N. Greenwood, *Intern. At. Energy Agency, Tech. Rept. Ser.* **50**, 163 (1966).
19. N. N. Greenwood, P. G. Perkins, and D. H. Wall, *Symp. Faraday Soc. No.* 1, 51 (1967).
20. H. A. Bent, *J. Inorg. Nucl. Chem.* **19**, 43 (1961); *Chem. Rev.* **61**, 275 (1961).
21. M. Cordey-Hayes, *J. Inorg. Nucl. Chem.* **26**, 2307 (1964).
22. A. Yu. Aleksandrov *et al.*, *Dokl. Akad. Nauk SSSR* **165**, 593 (1965).
23. See discussion by V. I. Gol'danskii *et al.* [2], p. 343.
24. V. V. Khrapov, Ph.D. Thesis, *Inst. Chem. Phys. Akad. Sci., USSR*, Moscow, 1965.
25. B. Gassenheimer and R. H. Herber, *Inorg. Chem.* **8**, 1120 (1969).
26. T. C. Gibb and N. N. Greenwood, *J. Chem. Soc.* **1966**, 43.
27. N. N. Greenwood and J. N. R. Ruddick, *J. Chem. Soc. A* **1967**, 1979.
28. T. A. Manuel and F. G. A. Stone, *Proc. Chem. Soc.* **1959**, 90; *J. Am. Chem. Soc.* **82**, 336 (1960).
29. M. D. Rausch and G. N. Schrautzer, *Chem. Ind. (London)* 957 (1959).
30. A. Nakamura and N. Hagihara, *Bull. Chem. Soc. Japan* **32**, 880 (1959).
31. F. A. Cotton, *J. Chem. Soc.* **1960**, 400.
32. B. Dickens and W. N. Lipscomb, *J. Am. Chem. Soc.* **83**, 489 (1961).
33. C. G. Kreiter *et al.*, *J. Am. Chem. Soc.* **88**, 3444 (1966).
34. F. A. Cotton, A. Davidson, and J. W. Faller, *J. Am. Chem. Soc.* **88**, 4507 (1966).
35. C. E. Keller, B. A. Shoulders, and R. Pettit, *J. Am. Chem. Soc.* **88**, 4760 (1966).
36. R. Grubbs, R. Breslow, R. H. Herber, and S. J. Lippard, *J. Am. Chem. Soc.* **89**, 6864 (1967).
37. F. Bonati and G. Wilkinson, *J. Chem. Soc.* **1964**, 79.
38. R. H. Herber, *Symp. Faraday Soc. No.* 1, 86 (1967).
39. J. E. O'Connor and E. R. Corey, *Inorg. Chem.* **6**, 969 (1967).
40. R. H. Herber and Y. Goscinny, *Inorg. Chem.* **7**, 1293 (1968).
41. M. A. Bush and P. Woodward, *J. Chem. Soc. A* **1967**, 1833.
42. W. Jetz and W. A. Graham, *J. Am. Chem. Soc.* **89**, 2773 (1967).
43. W. Jetz, P. Simons, J. Thompson, and W. A. G. Graham, *Inorg. Chem.* **5**, 2217 (1966).
44. H. R. H. Patil and W. A. G. Graham, *Inorg. Chem.* **5**, 1401 (1966).
45. R. H. Herber and H. -S. Cheng, *Inorg. Chem.* **8**, 2145 (1969).
46. H. -S. Cheng and R. H. Herber, *Inorg. Chem.* **9**, 1686 (1970).

Chapter 8

Mössbauer Spectroscopy and Physical Metallurgy

U. Gonser

Universität des Saarlandes
Saarbrücken, Germany

A decade ago Rudolf Mössbauer [1,2] observed that the emission and absorption of gamma rays can occur in a recoil-free fashion. This discovery led to a new scientific tool—the Mössbauer effect. The wide applicability of the Mössbauer effect has had a great impact on many disciplines in natural sciences over the last decade, including physical metallurgy. One might point out that the gain in information was not always a one-way street. The two fields of nuclear physics and metallurgy especially played a significant role in the development and application of Mössbauer spectroscopy. Knowledge of nuclear physics parameters, such as nuclear moments, were important in the understanding of the hyperfine pattern, and physical metallurgy considerations contributed to various practical aspects of this new technique.

Concerning the latter point, one might add a human factor often encountered in life: the outcome and results are featured and discussed in great length and detail; however, the frustrations and difficulties and "witchcraft" involved in preparing appropriate materials (sources and absorbers) are often neglected.

Some areas where Mössbauer spectroscopy profited from physical metallurgy, and also the reverse, where metallurgy gained knowledge from Mössbauer spectroscopy, might be written schematically as shown in Table 1. Following this scheme we have divided this chapter into two parts: Mössbauer application to physical metallurgy and physical metallurgical considerations in Mössbauer spectroscopy. It should be emphasized that the selected systems and examples are somewhat arbitrarily chosen, and the spectra shown should be considered as typical representatives in the field.

**Table 1. Schematic Representation of Interaction between Mössbauer
Effect and Physical Metallurgy**

Mössbauer effect	⇄	Physical metallurgy
Preparation of source and absorber		Magnetic structures
Preferential orientation (texture)		Néel and Curie point determination
Magnetic domain structure		Phase identification
Radiation damage in the source		Near-neighbor configuration
Lattice defects		Ordered alloys
Standards		Precipitation processes
		Thin films
		Small particles
		Lattice dynamics
		Lattice defects

The Mössbauer effect has been demonstrated in about 65 isotopes [3].
Some elements have various isotopes (up to six in the case of Gd) useful for
Mössbauer spectroscopy. In many cases, it was not easy to find a resonance
effect because of unfavorable nuclear and solid-state parameters involved:
high gamma-ray energy of the excited state (low recoil-free fraction), high
internal conversion coefficient, short lifetime of the parent isotope, difficulties in preparing appropriate source and absorbers, etc. In most of the
experiments at least one of the resonators (source or absorber) was in the
metallic state.

Certainly in one respect nature showed a friendly attitude: the isotope
^{57}Fe, which is most favorable for Mössbauer spectroscopy, belongs to the
element most important in physical metallurgy. Also, all the important
effects, such as isomer shift, nuclear Zeeman effect, quadrupole splitting,
gravitational red shift, temperature shift, etc., were first demonstrated with
this isotope. Although the natural abundance of ^{57}Fe is only 2.19%, in
many applications isotopical enrichment is unnecessary. According to the
Mössbauer Effect Data Index by Muir et al. [3], more than 53% of the
reported work on the Mössbauer effect deals with this isotope.

The history of the Mössbauer effect sketched in Frauenfelder's book
[4] is divided into the prehistoric time (remarkable, insofar as that the
discovery was not made before 1958), the early, middle, and late iridium age
(Mössbauer found the effect with ^{191}Ir), and the iron age (discovery of the
effect in ^{57}Fe) with the appropriate remark, "wow." These introductory
remarks might be considered as an excuse that we concern ourselves in this
chapter mainly with the isotope ^{57}Fe.

1. MÖSSBAUER APPLICATION TO PHYSICAL METALLURGY

A significant application of the Mössbauer effect to physical metallurgy
is connected with (a) the qualitative and quantitative identification of

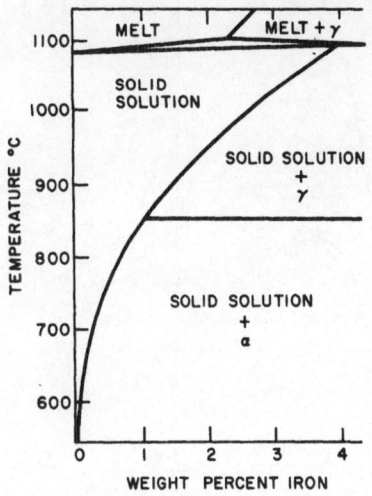

Figure 1. The Cu–Fe phase diagram on the copper-rich side.

various alloy phases and (b) its ability to measure specific properties of the phases. We want first to present a somewhat simple case, that is, the copper-rich Cu–Fe system [5].

1.1. Precipitation in the Cu-Fe System

The Cu–Fe phase diagram on the copper side is shown in Figure 1. According to this phase diagram, the solubility of iron in copper is about 4 wt % Fe ($\approx$ 4.5 at. % Fe) at the melting point and drops sharply with temperature. At 500°C the solubility of Fe in Cu is extremely small. Supersaturated solutions can be obtained by fast quenching techniques. Subsequent annealing at elevated temperatures but below the solubility line causes the Fe to precipitate. Even below the temperature of the $\gamma \rightarrow \alpha$ transition, the Fe precipitates coherently with the fcc copper matrix as fcc γ-Fe. In Figure 2a the Mössbauer spectrum of a 0.6 at. % Fe–Cu absorber is shown after solution annealing at 875°C and quenching in water [6]. The spectrum can be interpreted in terms of the presence of at least two phases: the absorption component to the right (I) due to the Fe in solution and the component to the left (II) due to "γ-Fe precipitates." The isomer shift of the two phases makes the decomposition possible. The curves drawn in the figure are least-squares Lorentzian single line curves for the two components adjusted to produce envelopes of best fit. Some of the γ-Fe precipitates in the quenched sample might be extremely small, consisting effectively of only iron clusters (dimer, trimer, etc.). As a result of the nearest neighbor interaction, the local cubic symmetry is removed, and a quadrupole interaction is expected. The rather bad fit on the positive velocity side might be

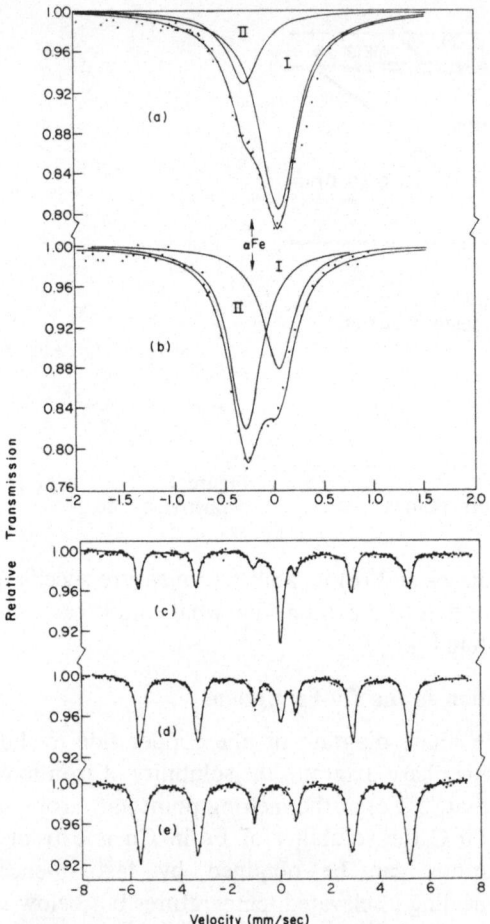

Figure 2. Mössbauer transmission spectra for
Cu–Fe absorbers at 80°K. (a) 0.6 at.% Fe
sample solution annealed at 875°C and quenched (decomposition I and II are explained in the
text); (b) 0.6 at.% Fe sample after annealing
5 min at 600°C; (c) 0.6 at.% Fe sample annealed 72 h at 600°C and cold-rolled; (d) 3.5 at.%
Fe sample annealed 27 h at 650°C and cold-
rolled; and (e) pure α-Fe foil at 80°K.

seen as an indication of one of the quadrupole lines, while the other is
hidden under line II. An analysis in terms of three components (Fe in
solution, γ-Fe precipitates, and small atomic clusters) is rather difficult
because of insufficient resolution and statistical uncertainties. The deviation
from a single line spectrum indicates that even a fast quench does not prevent

the formation of precipitates of iron atoms and demonstrates the difficulties of completely retaining the iron in a supersaturated solid solution.

The copper-rich Cu–Fe system resembles the copper-rich Cu–Co system, especially in regard to the solubility. Consequently, one might expect precipitation of Co atoms to occur in the latter in a similar fashion. However, the Co precipitates—at least above a minimum particle size—are ferromagnetic in contrast to the γ-Fe precipitates, which are paramagnetic at room temperature. The Cu–Co system is of interest to Mössbauer spectroscopy because ^{57}Co diffused into copper is frequently used as a single line source (copper source). Although the absolute amount of ^{57}Co is mostly rather small ($<10^{-7}$ atomic fraction), one has to realize that one is dealing with the supersaturated, thermodynamically metastable system.

In the earlier work with copper sources, we observed an interesting effect. The originally single line source developed, over the period of a year, a resonance contribution in the wings that could be identified as a six-line Zeeman pattern. The ^{57}Co available at that time came with a relatively high amount of inactive ^{59}Co (^{59}Co/^{57}Co$\approx$100). Over long periods of time, the few vacancies present in the material—making about 1 jump per minute at room temperature—lead to cluster formations and ferromagnetic cobalt precipitates.

The quenched sample (0.6 at. % Fe) used for the spectrum in Figure 2a was subsequently annealed for 5 min at 600°C. The spectrum obtained after this treatment is shown in Figure 2b. The increase in intensity of line II indicates the growth of the γ-Fe precipitates while the matrix becomes diluted on iron (decrease of the intensity of line I). With longer annealing time a larger fraction of Fe precipitates out of the matrix, and the γ-Fe precipitates increase in size.

Iron can be retained in the fcc structure at low temperature by precipitating iron from a supersaturated solution, for instance, in the Cu–Fe system as discussed, or by alloying iron with certain elements, such as nickel and cobalt, which widen the stability range of the γ-phase. If fcc iron is measured by Mössbauer spectroscopy, a broadening of the line occurs at 38°K for stainless steel and at about 55–67°K, depending on the particle size, for γ-Fe precipitates in a copper matrix [7]. In Figure 3 the transmission spectra for a 304 stainless steel sample just above and below the transition region are shown. The broadening was interpreted as a magnetic transition from the paramagnetic to the antiferromagnetic state of γ-Fe. The internal field is rather small in this case, ~ 25 kOe; thus, the lines of the Zeeman pattern remained unresolved. In order to determine the Néel temperature, the following simple method was employed, which has proven to be convenient in similar cases. The count rate at zero velocity was measured as a function of temperature as is shown in Figure 4. The onset of the splitting

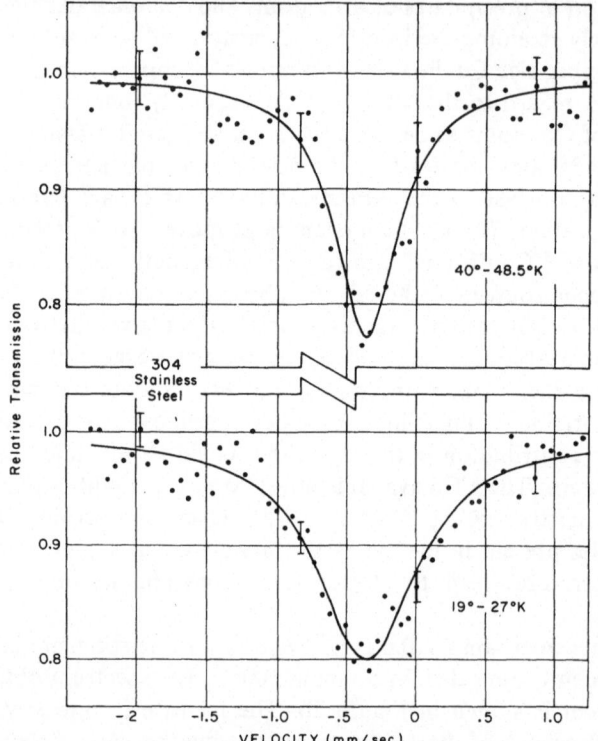

Figure 3. Mössbauer transmission spectra of a 304 stainless
steel sample just above and below the transition temperature.

(unresolved) that lowers the effective Mössbauer cross section is measured
and equated to the Néel temperature in the material.

γ-Fe (fcc) precipitates in a copper matrix will transform to thermodynamically stable bcc α-Fe on plastic deformation. This can easily be

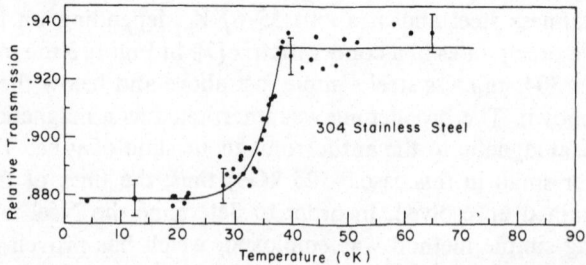

Figure 4. Count rate at zero velocity of a 304 stainless
steel sample as a function of temperature.

followed by the appearance of the characteristic six Zeeman lines of α-Fe. However, a certain minimum α-Fe particle size is required before the Zeeman effect can be seen. If the particles are very small, paramagnetic spectra or spectra typical for superparamagnetic behavior are expected. Figure 4 shows two samples (0.6 at. % Fe and 3.5 at. % Fe) that have been solution-annealed at high temperature, quenched, and then annealed at 600°C for 72 h (spectrum c) and at 650°C for 27 h (spectrum d) and subsequently cold-rolled (~50% reduction in thickness). The ratio of the iron present in the ferromagnetic α-phase (corresponding to the six-line pattern) to the iron which is still in solution, in γ-Fe precipitates, and possibly in α-Fe precipitates which are too small to show ferromagnetic behavior (corresponding to the central line), can be obtained from the relative line intensities.

Ron et al. [8] made the observation that the area under the ferromagnetic six-line spectra of α-Fe precipitates decreases markedly by applying a transverse magnetic field in the order of several thousand oersteds. This decrease was explained as a change of the recoil-free fraction of the iron in the particles. However, it was pointed out [9] that this interpretation is incorrect: in applying a field, one has to consider the polarization effects associated with the Mössbauer absorption that changes the relative as well as the absolute line intensities.

The early state of processes leading to precipitation is characterized by a deviation from random distribution of a supersaturated solid solution. Whether precipitation in a system occurs by nucleation and growth of a new phase or by spinodal decomposition can be decided in favorable cases by Mössbauer spectroscopy. Nagarajan and Flinn [10] have suggested that the decomposition of the solid solution of Cu–Fe–Ni occurs by nucleation and growth process and not—as has been widely believed—by a spinodal mechanism.

1.2. Phase Transition in Stainless Steel

In some types of stainless steel a certain fraction of the material is transformed to a magnetic ferrite phase under extreme mechanical deformation. A 302 stainless steel sample, 0.025 mm thick, has been subjected to a 283 kbar explosive shock treatment [11]. The Mössbauer transmission spectra before and after the treatment are shown in Figure 5. The spectra indicate that about 50% of the steel has been transformed to the martensite phase.

Up to now the experimental metallurgical studies involving Mössbauer spectroscopy have been made in the conventional transmission mode where the material is a thin foil or powder. The fabrication of thin foils is tedious and often impractical, especially if the resonance technique is used as a routine analytical tool. One can foresee that in the future the scattering technique (see Chapter 2) will be employed more and more for metallurgical

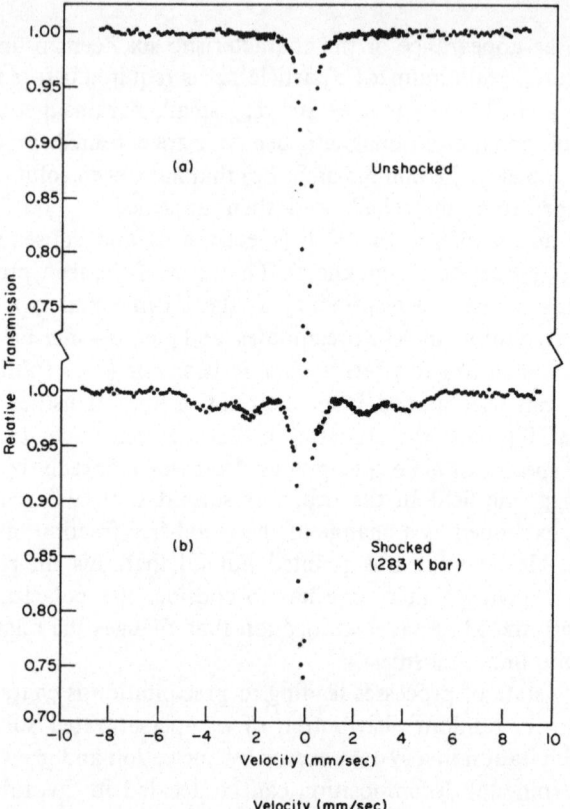

Figure 5. Mössbauer transition spectra of a 302 stainless steel sample (a) before and (b) after 283 kbar explosive shock treatment.

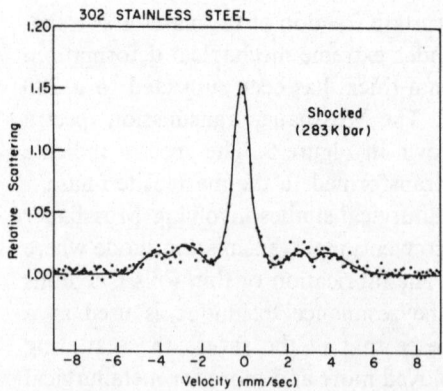

Figure 6. Mössbauer scattering spectrum of a 302 stainless steel sample after 283 kbar explosive treatment.

problems. The scattering spectrum of the shock-treated 302 stainless steel sample is shown in Figure 6. It is evident that the two spectra, Figures 5 and 6, of the same sample taken in the transmission and scattering mode, exhibit the same features.

1.3. Internal Oxidation Studies

The internal oxidation of Fe to FeO, and subsequently to Fe_3O_4 and $CuFeO_2$, which occurs by heat treatments of Cu–Fe samples in a relatively low partial pressure atmosphere of O_2, can be conveniently followed by Mössbauer spectroscopy. Although FeO has the cubic sodium chloride structure, it exhibits, at room temperature, a quadrupole split spectrum with an isomer shift characteristic of divalent Fe. The removal of the local cubic symmetry in the Fe-deficient compound ($Fe_{0.87-0.95}O$) has been thought to be the reason for the presence of the electric field gradient at the site of the nuclei [12].

Fe_3O_4 shows, at room temperature, two superimposed six-line Zeeman spectra [13,14]. The two patterns are due to Fe^{3+} at tetrahedral sites (H_n $=492\pm5$ kOe) and Fe^{2+} and Fe^{3+}—with rapid charge exchange—at octahedral sites ($H_n=464\pm5$ kOe).

The terminal oxidation product has been identified as $CuFeO_2$, which occurs naturally as the mineral delafossite and in copper smelter furnace slags. The isomer shift observed is characteristic of trivalent iron which provides conclusive evidence of the ionic charge state of delafossite: $Cu^+Fe^{3+}O_2$ (not $Cu^{2+}Fe^{2+}O_2$). A magnetic transition was found with a Néel temperature of 19 °K [15].

In summary, the following reactions have been measured in the Cu–Fe system

$$Fe_{solution} \xrightarrow{\text{annealing}} \gamma'\text{-Fe (precipitation)} \tag{1}$$

$$\gamma\text{-Fe} \xrightarrow{\text{deformation}} \alpha\text{-Fe (phase transformation)} \tag{2}$$

$$Fe + \tfrac{1}{2}O_2 \xrightarrow{\text{annealing}} FeO \text{ (internal oxidation)} \tag{3}$$

$$3FeO + \tfrac{1}{2}O_2 \xrightarrow{\text{annealing}} Fe_3O_4 \text{ (internal oxidation)} \tag{4}$$

$$Fe_3O_4 + 3Cu + O_2 \xrightarrow{\text{annealing}} 3CuFeO_2 \text{ (internal oxidation)} \tag{5}$$

1.4. Magnetic Properties in Au–Fe Alloys

Transition metals dissolved in noble metals show some peculiarities in their magnetic properties. Various techniques have been used to obtain information about the nature, origin, and magnetic structure of these alloys. In the investigations of magnetically dilute alloys, the Mössbauer effect is particularly useful [16,17] because one can observe the distribution of alignments of the atomic spins, while in the conventional methods (magnetic susceptibility, magnetization, remanence), only the average alignment is observed.

In metallic alloys the Mössbauer effect measures the time-average magnetic field at the site of the nucleus $<H_n>$ during the lifetime of the excited state. For an isotope in a magnetic environment, $<H_n>$ and the time average of the atomic spin $<S>$ are normally found to be nearly proportional ($<H_n> \propto <S>$) so that one can derive from the $<H_n>$ behavior the $<S>$ properties. Above the Curie temperature (paramagnetic state), the relaxation time of the atomic spins is short compared to the lifetime of the excited state and the nuclear Zeeman effect in zero applied field averages to zero. Below the Curie temperature the spontaneous align-ment of the atomic spin produces an internal magnetic field. If this field is sufficiently large, a resolved hyperfine spectrum is observed.

Gold–iron alloys are particularly interesting because of the large solubility of iron in the fcc phase (75 at.% at 1168°C and about 16 at.% at 400°C). While in the Cu–Fe alloys it is hard to avoid precipitation even by fast quenching, in the Au–Fe alloys it is difficult to produce precipitates, at least in the concentration range up to 25 at.%.

The spectra above and below the magnetic transition of an Au-19.5-at.% Fe sample are shown in Figure 7. For iron in a cubic environment, one expects a single line in the paramagnetic state and a six-line spectra in the magnetic state. The deviations from this behavior indicate the Fe–Fe nearest-neighbor interaction. The transition temperatures of Au–Fe random solid solutions measured in a similar way as described in the case of fcc stainless steel (Figure 4) as a function of concentration along with measurements using other techniques are plotted in Figure 8. The transition temperature increases with increasing iron concentration. However, the slope changes at

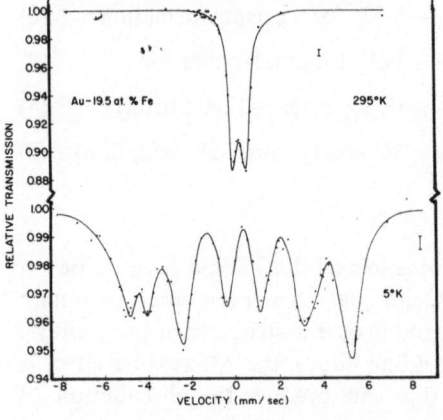

Figure 7. Mössbauer transition spec-tra of an Au–19.5 at.% Fe sample at 295°K and 5°K.

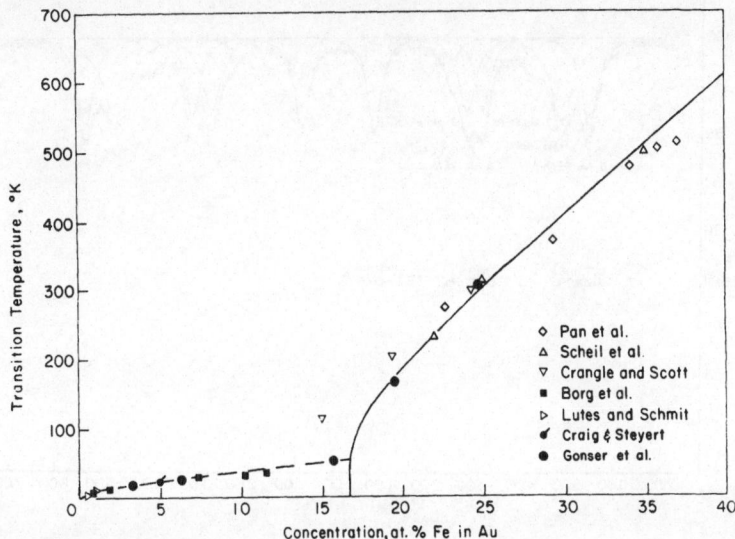

Figure 8.   Magnetic transition temperature in the Au–Fe alloy system as a function of iron concentration.

a critical concentration of ~ 16 at. % Fe. By applying an external magnetic field (up to 55 kOe) to alloys ≲ 16 at. %, Fe, the atomic spins do not align with the field (the $\Delta m = 0$ lines remain strong), indicating spacially random spin orientation. In the high-concentration alloys (≳ 16 at. % Fe), the $\Delta m = 0$ lines disappear, indicating alignment of the spin with the applied external magnetic field [18,19]. Thus, the two branches, up to 16 at. % Fe, can be associated with antiferromagnetic and ferromagnetic behavior, respectively

1.5.   Near-Neighbor Interaction in Substitutional α-Fe–Mo Alloys

An impurity atom incorporated in a magnetic matrix will have an effect on the charge distribution and spin density in its environment. As a result of this, the Mössbauer spectra become more complex. It is often possible to unscramble the pattern and correlate the various superimposed spectra to the various shells (nearest neighbor, next nearest neighbor, etc.) surrounding the impurity atom. The effects on the hyperfine interaction of ^{57}Fe nuclei have been distinguished up to six neighbor shells away. In the analysis, random distribution of the solute atom is mostly assumed. The substitutional systems of α-Fe with various amounts of V, Cr, Mo, Al, Ti, Si, Mn, Ga, Sn, and Be have been investigated, especially by Stearns [20] and others [21–26]. As an example, some results in the substitutional α-Fe–Mo system obtained by Marcus *et al.* [26] are presented.

Figure 9 shows the Mössbauer spectrum of an Fe–6 at. % Mo sample

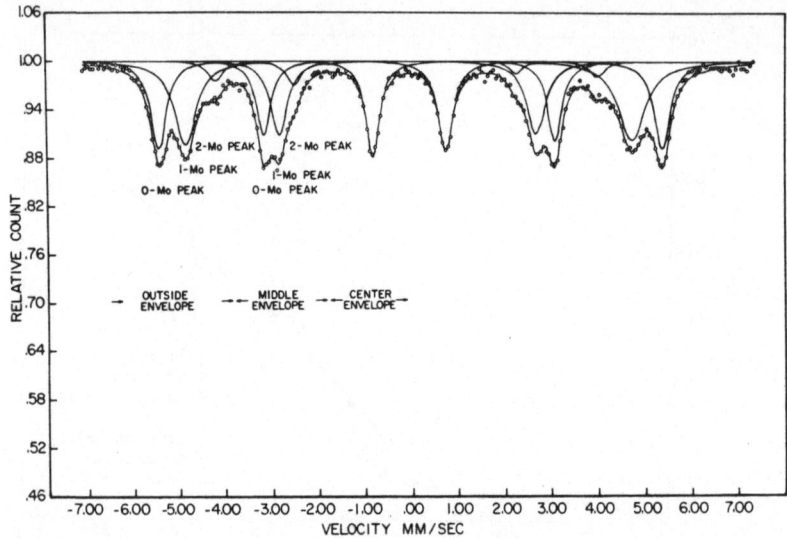

Figure 9.   Mössbauer transition spectrum of an Fe–6 at. % Mo alloy quenched from 1050°C. The experimental curve was decomposed into its component using a 14-nearest-neighbor model (after Marcus *et al.* [26]).

solution annealed at 1050°C and quenched. The spectrum can be decomposed into three Zeeman patterns. Because the first two coordinate shells (14 nearest neighbors) in the α-Fe bcc structure have nearly the same radii, it was assumed that a solute atom in both shells would contribute the same effect to the charge and spin density on the iron atom. This assumption seems in many solute systems verified; however, it may not be valid in some other cases. On the basis of the 14-nearest-neighbor model, the three superimposed spectra were interpreted in terms of iron with zero neighbors ($H_n = 335 \pm 1$ kOe), iron with one molybdenum neighbor ($H_n = 296 \pm 1$ kOe), and iron with two molybdenum neighbors ($H_n = 255 \pm 1$ kOe). The decomposition of the spectrum is shown in Figure 9.

1.6.   Near-Neighbor Interaction in Interstitial Fe–C Alloys

Carbon, nitrogen, and hydrogen, and probably oxygen and boron dissolve as interstitial atoms in α-Fe and γ-Fe. The distortion of the lattice caused by these interstitial atoms has been studied for many years by various techniques. Here we want to focus our attention on the Fe–C system [27]. Of great interest is the hope of extracting data and a deeper understanding of the hardening mechanism, important from a technological point of view. Carbon enters fcc γ-Fe (austenite) in interstitial lattice sites with octahedral symmetry. The presence of carbon atoms can be observed by the expansion

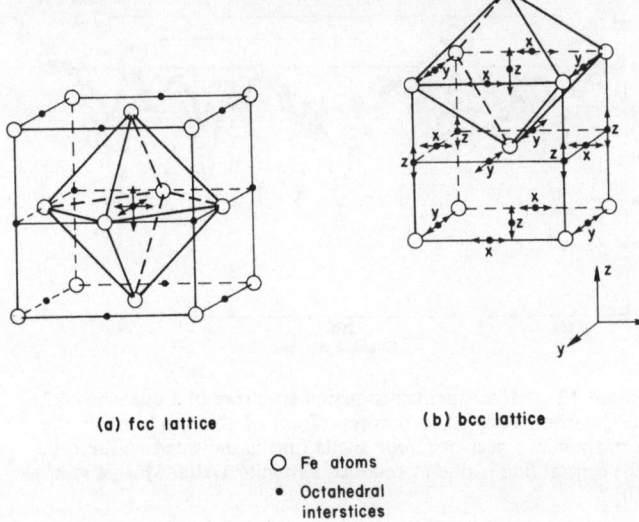

(a) fcc lattice   (b) bcc lattice

O Fe atoms

• Octahedral
  interstices

Figure 10. Unit cells (fcc and bcc) drawn approximately in proportion to the lattice parameters of γ-Fe and α-Fe. The solid circles indicate the positions of the octahedral interstices. Arrows show axes of four-fold symmetry.

of the lattice. It is believed that a cubic symmetric distortion occurs, as indicated in Figure 10a. In the bcc α-Fe the corresponding octahedra are not regular but are shortened along one of the x, y, and z axes as indicated by the arrows and letters in Figure 10b. Each octahedron has only one four-fold symmetry axis in α-Fe, while the regular octahedron site in γ-Fe has three four-fold axes through the opposite pairs of atoms. The oriented localized octahedral distortion in bcc α-Fe was called by Cohen [27] "dis-

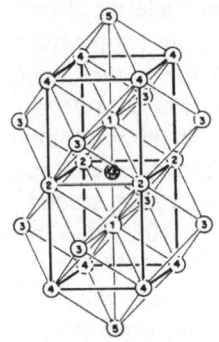

⊕ Carbon atom
O Iron atom

Figure 11. Crystal structure of bct iron–carbon martensite around a carbon atom. Numbers indicate near-neighbor shells.

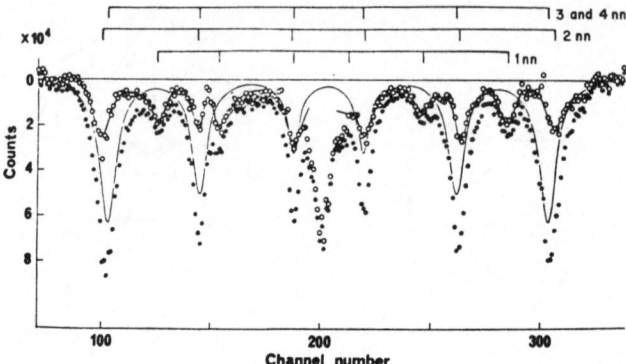

Figure 12. Mössbauer transmission spectrum of a quenched 4.2 at.% carbon alloy. The decomposition of the spectrum to the corresponding near-neighbor shells (nn) is indicated at the top. The central line is due to retained austenite. (After Moriya *et al.* [29].)

tortion dipole." When carbon–austenite transforms by a diffusionless process (martensite transformation) to martensite, the resulting structure is no longer bcc but body-centered tetragonal, bct. The carbon atoms are located in the martensite structure between two iron atoms which are aligned along one expanded axis, the tetragonal c axis. X-ray measurements have shown that the c/a ratio of martensite increases nearly linearly with the carbon content. In Figure 11 the iron sites in the bct martensite structure around an interstitial carbon atom are shown. The numbers indicate the neighboring shells according to the distance to the carbon atom at the origin.

Some interesting Mössbauer spectroscopy measurements in the iron–carbon austenite and martensite phases have been obtained [28–31]. A 4.2 at.% carbon alloy quenched in water had been used by Moriya *et al.* [29]. The Mössbauer spectrum shown in Figure 12 was interpreted in terms of two phases: (a) martensite exhibiting a hyperfine spectrum at room temperature and (b) paramagnetic retained austenite exhibiting a single line in the center of the spectrum. The martensite phase can be analyzed further by considering the various possible nearest-neighbor (nn) environments of the randomly distributed carbon atoms. For example, for the 4.2 at.% carbon concentration, the probability of finding configurations with first-nn Fe's, second-nn Fe's, third-nn Fe's, etc., are 8.2%, 14.5%, 22%, 16%, 3%, 10%, etc. The probabilities have to match the intensities in the analysis of the Mössbauer spectrum. Furthermore, the reasonable assumption is made that the first- and second-nn Fe's are strongly effected by the carbon atom, while the Fe atoms in the outer shells will exhibit spectra resembling more that of α-Fe.

The Zeeman components originated from martensite can be decom-

Table 2. The Internal Fields, Isomer Shifts, and Electric Quadrupole Interactions for Different Iron Sites Near the Carbon Atoms in the Fe–4.2 At.% C Martensite Structure[a]

	Population, %	Internal field, H_n, kOe	Isomer shift, mm/sec	Quadrupole splitting, mm/sec
First-nn Fe	8.2	265±2	−0.03±0.05	0.13±0.05
Second-nn Fe	14.5	342±2	0.02±0.05	−0.02±0.05
Third-, fourth-nn Fe	38	334±2	0.01±0.05	0.01±0.05
Pure iron		330	0	0

[a] After Moriya et al. [29]. δ relative to iron.

posed to the corresponding nearest neighbor configurations (1 nn, 2 nn, 3 nn, and 4 nn) as indicated above the spectrum. Actually, the spectrum is simplified by subtracting an appropriate intensity (39%) of a pure iron component corresponding to nn ≥ 5 which is considered undistinguishable from the α-Fe spectrum. The Mössbauer spectrum for the different iron sites are listed in Table 2. As can be seen from this table, the isomer shift (vs α-Fe) and the quadrupole interaction (except for the first-nn Fe's) are small but the hyperfine fields are appreciably affected by the presence of the carbon atoms. In Figure 13 the internal fields of the various nn Fe sites are plotted as a function of the iron–carbon distance. It is interesting to note that the first nearest neighbor has a much smaller internal field than pure α-Fe, while the next nearest-neighbor shells have larger fields than α-Fe.

We have seen that the Mössbauer effect can distinguish various simultaneously present phases, and within these phases, various environments. Considering this fact, it seems natural that the effect became an important tool in the investigation of the various phases found in steel. The publi-

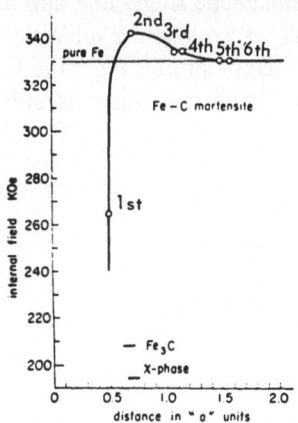

Figure 13. The internal magnetic field of the different nn sites as a function of the carbon–iron distance. (After Moriya et al. [29].)

cations on this topic have mushroomed over the last two years [28–41]. Of course, one has to keep in mind that it is often very difficult to unravel the spectra because of the variety of substitutional and interstitial impurities and the presence of the different simultaneously present phases: martensite, austenite, cementite, ε-phase, and others.

1.7. Order-Disorder in FeAl and Fe₃Al

In FeAl (CsCl structure) no magnetic moment is associated with the iron atom. The ordered alloy exhibits a single-line Mössbauer spectrum at room temperature. However, nonstoichiometric and crushed FeAl alloys show a single line and a hyperfine pattern simultaneously. These observations were interpreted by considering the nearest neighbor configuration. The plastic deformation will form a large number of antiphase boundaries, and thus many Fe–Fe nearest neighbor interactions. The effect of the antiphase boundaries is to produce groups of moment-bearing atoms large enough to have ferromagnetic behavior [42,43].

Fe₃Al and Fe₃Si are of special interest and have been intensively investigated by various methods. The structure shown in Figure 14 consists of two interpenetrating simple cubic sublattices: sublattice A containing iron atoms (with four iron and four aluminum nearest-neighbor atoms) and sublattice D containing, alternatingly, iron and aluminum atoms (with eight nearest-neighbor iron atoms). From neutron scattering experiments at room temperature [44], the magnetic moments at the A and D sites are 1.46 μ_B and 2.14 μ_B, respectively. The internal magnetic fields found by Mössbauer spectroscopy for the A and D sites are 210 and 294 kOe, respectively [45,46]. At lower temperature—up to about 0.9 T_C—the variation of the reduced internal field with reduced temperature was almost identical for both sublattices. In the temperature range of 500–640°C, Cser *et al.* [47] observed the simultaneous appearance of a paramagnetic single line and a hyperfine spectrum. This was explained on the basis of two distinct sublattice magnetic ordering temperatures. In the indicated temperature range, the D sublattice becomes magnetically ordered while the A sublattice is still magnetically disordered.

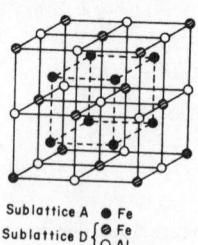

Sublattice A   ● Fe
Sublattice D $\left\{ \begin{array}{l} ⊕ \text{ Fe} \\ ○ \text{ Al} \end{array} \right.$

Figure 14.   Structure of Fe₃Al.

The onset of the crystallographic ordering of the D sublattice, change from the pseudoorder of type B2 to DO_3-type ordering, is accompanied by a pronounced discontinuity in the internal field and isomer shift. It was pointed out by Cser et al. [47] that the composition of the second coordination shell of the D sublattice atoms changes markedly during the ordering. That is, instead of containing on the average three iron and three aluminum atoms, it will have six aluminum atoms. At higher temperature (800–850°C) a change in the isomer shift occurs, indicating the transition from the disordered to the B2-type pseudo-order form. The determination of the magnetic and crystallographic ordering parameters in the Fe_3Al system shows the remarkable potential of the Mössbauer effect. A similar investigation on the order–disorder has been carried out in the Fe–Rh system by Shirane et al. [48].

1.8. Thin Films and Superparamagnetism

Lee et al. [49] showed that Mössbauer spectroscopy is well suited for the investigation of ultrathin foils, and the question could be answered: at what thickness does the saturation magnetization and Curie temperature of ferromagnetic films depart from bulk value? Thin Fe films with an average thickness D between 1 and 1220 Å were obtained by vacuum evaporation. A high vacuum prevented the oxidation of the films. The measurements were made from 4 to 773°K. With $D \geq 45$ Å, the hyperfine spectrum cannot be differentiated from pure bulk α-Fe. The intensity ratio of the pattern 3:4:1:1:4:3 indicates that the magnetization lies in the plane of the foil. With $D \leq 45$ Å, the spectrum taken at room temperature shows line broadening, which was attributed to superparamagnetic behavior (see below). In the range $5 \leq D \leq 15$ Å, a decrease in the internal field is observed at room temperature, and also the Curie temperature decreases with thinner samples. Finally, with extremely thin samples $D \leq 5$ Å, the Zeeman pattern collapses at room temperature because the internal field averages to zero on the time scale of Mössbauer spectroscopy. The removal of the cubic symmetry close to the surface is probably the reason for the appearance of two lines (quadrupole interaction).

Thin films and single domain particles that are sufficiently small in size show the phenomena of superparamagnetism: the thermally activated fluctuation of magnetization in the different easy directions [50,51]. The fluctuation processes of the magnetization vector of a particle with average volume V and an anisotropy constant K can be described by the approximate expression for the relaxation time $\tau = \tau_0 \exp (KV/kT)$, where τ_0 is the angular precession frequency (in the order of 10^{-9} sec), k is the Boltzmann constant, and T is the absolute temperature. The characteristic time scale of measurements with conventional methods is of the order of seconds but very small

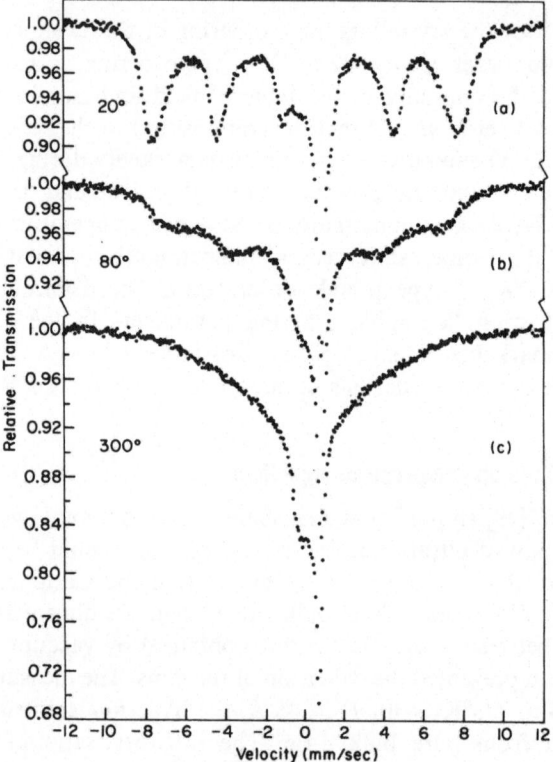

Figure 15.   Transmission Mössbauer spectra for MgFe₂O₄
precipitates in MgO at (a) 20°K, (b) 80°K, and (c) 300°K.

in Mössbauer spectroscopy, in fact, of the order of $\tau_m \sim h/\Delta E$, where ΔE is
the total hyperfine splitting. When $\tau_m \sim \tau$, the hyperfine split spectrum collapses as exemplified on MgFe₂O₄ small precipitates in Figure 15. The
single absorption line at $\simeq +1$ mm/sec is due to Fe^{2+}, which remained in
solution. If K is known, the particle size can be deduced from these spectra
taken at various temperatures. Nasu et al. [52] found superparamagnetism
in small cobalt precipitates in a copper matrix.

1.9.   Mössbauer Effect as a Nondestructive Analytical Tool

Mössbauer spectroscopy is a nondestructive method of investigation.
Therefore, it seems appealing to use the Mössbauer spectra as a quantitative
analytical tool to measure various phases simultaneously present. In such
analyses one has to make corrections for thickness and preferred crystallographic orientation of crystallites (texture), the recoil-free fractions of the

resonating isotopes have to be known, and the site populations and polarization effects have to be taken into account.

When Mössbauer spectroscopy is used as a standard analytical method of multiphase systems, the recently advanced stripping technique is very helpful [53]. This analytical technique involves reference spectra in least-squares-fit Lorentzian representations that are subtracted by appropriate amounts from the measured spectrum. The iteration procedure is continued until the residuals become satisfactorily small, or in other words, every part of the resonance spectrum has been quantitatively assigned to a specific phase. The technique has been proven particularly useful in analyzing meteorite spectra where as many as six iron containing phases have been distinguished [54,55].

The significance of developing the resonance scattering technique has been pointed out previously.

2. PHYSICAL METALLURGY CONSIDERATIONS CONCERNING THE MÖSSBAUER EFFECT

In Mössbauer spectroscopy the source spectrum is related to the absorber spectrum by the Doppler motion. Therefore, one of the two has to be known in the determination of the other. In special experiments, for instance, if one is interested in the dispersion associated with polarized gamma rays in a transmitter (Faraday effect and magnetic double refraction) [56–58], both the source and the absorber spectra have to be known. In most cases measurements are made in the absorber mode in conjunction with a single-line-source spectrum. In fact, the source line position is mostly taken as a standard reference position. This is often unsatisfactory, especially when various spectra taken with different sources—and thus different isomer shifts—are compared. It would be desirable to have a unique reference standard acceptable for physicists, chemists, metallurgists, etc. for each isotope. Unfortunately, the Mössbauer community is as split as their spectra on the choice of standard sources and absorbers. Most desirable seems a standard for ^{57}Fe with its lion's share in the field. Sodium nitroprusside, $Na_2[Fe(CN)_5NO] \cdot 2H_2O$ has been suggested as a standard [59, 60]. While the chemists seem to favor sodium nitroprusside, the physicists and metallurgists are more inclined to use α-Fe as a standard.

For the sources, mostly metallic matrices are chosen in which the parent or excited isotopes are diffused, implanted [61,62], Coulomb-excited [63,64], or produced by nuclear reaction [65,66]. Some of the unique advantages of metal source are as follows.

(a) Most metals have high coordination symmetry. In nonmagnetic cubic metals (fcc or bcc) with low impurity concentrations, single-line emission spectra are produced.

(b) The effective Debye–Waller factor at room temperature is relatively high for most metals.

(c) Electronic relaxation processes are extremely fast. Thus, localized charge states as found in insulators resulting from the foregoing decay, existing over long time periods compared to the lifetime of the excited states [67,68], do not exist in metals.

(d) In sources, lattice defects are sometimes created by the recoil of the processes leading to the excited nuclear resonance level: α and β decay [69,70], Coulomb excitation [63,64], implantation [61,62], and nuclear reaction processes [65,66]. These defects are intrinsically associated with the resonance isotope. In metals, the nature of the defects and their annealing are comparatively better understood than the defects created in insulators and semiconductors.

(e) Polarized gamma rays can be produced by magnetizing ferromagnetic sources, for instance, α-Fe.

(f) Most metals can be fabricated as thin foils; electroplating of the desired isotope is in most cases a straightforward procedure, and a uniform distribution of the isotope in the metal matrix can be achieved by an appropriate heat treatment. Stephen [71] gives a detailed description of the electrolytic methods and annealing treatments for metallic Mössbauer sources. Mössbauer sources are commercially available from various suppliers.

In general, one will recognize that the Mössbauer spectral lines in early papers were relatively broad and, consequently, the resolution was poor. This fact can be partially attributed to imperfections in the drive systems. That is, the count rate was not measured at a truly constant Doppler velocity. On the other side, some broadening can be understood in metallurgical terms. As mentioned before, a high impurity concentration (^{59}Co or Fe) or improper heat treatment and quenching rates might cause precipitation in ^{57}Co–Cu sources, and consequently broadening. Similarly, the distribution of the small isomer shifts and quadrupole splittings, in the manifold of environments in which the ^{57}Co finds itself in stainless steel sources, can be responsible for the relatively broad line. This is the reason that stainless steel sources, which are relatively easy to fabricate and were commonly used in the early days of ^{57}Fe Mössbauer spectroscopy, are no longer popular. One may point out that each of the commonly used unsplit ^{57}Co sources has at least one major drawback:

(a) ^{57}Co–Cu sources—possible precipitation of the supersaturated solution.

(b) ^{57}Co stainless steel—broad lines and self-absorption.

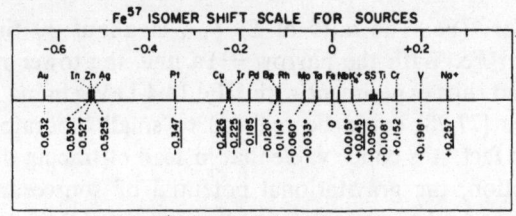

Figure 16. ^{57}Fe isomer-shift scale for sources relative to metallic α-Fe, both at room temperature: K_4^+ = $K_4Fe(CN)_6\cdot3H_2O$; $Na^+ = Na_2[Fe(CN)_5NO]\cdot2H_2O$; SS=stainless steel (300 series). Velocity is defined as positive for approaching relative motion between source and absorber. Thus, the source with the more negative shift has the larger nuclear energy level difference and hence the smaller electron density at the nucleus. After Muir et al. [3].)

(c) ^{57}Co in Pd—relatively high electronic absorption and interference of the K x rays.

(d) ^{57}Co in Pt—high electronic absorption interference of the L x rays.

(e) ^{57}Co–Cr—difficulty in fabricating foils. When using thick Cr sources the diffusion parameters have to be carefully selected so that ^{57}Co is distributed relatively homogeneously in the surface layer and the gamma rays are not attenuated too much by the matrix.

The isomer shifts in mm/sec from measurements made at room temperature with different sources (or referred to different absorbers) relative to metallic α-Fe are shown in Figure 16 [3].

^{119}Sn incorporated in metallic white tin or in SnO_2 were common sources in earlier work with this isotope. The observed Mössbauer resonance lines were rather broad and deviated from Lorentzian shape due to an unresolved quadrupole interaction in these noncubic lattices of white tin and SnO_2. Furthermore, theoretical and experimental measurements of the recoil-free fraction in single crystals of white tin have shown a considerable anisotropy [72–74]. Thus, the rolling or recrystallization texture will influence the Mössbauer spectrum. Such irreproducible parameters are certainly not acceptable in precision measurements. In search of new ^{119}Sn sources [75], the following matrices were suggested and are now in use: Mg_2Sn, Pd–Sn, Ag–Sn, and $BaSnO_3$. The Mg_2Sn sources have to be sealed or kept in a dry atmosphere; otherwise, they will decompose.

At present ^{181}Ta, with an excited state of $E\gamma=6.25$ keV, seems to be one of the most interesting isotopes in Mössbauer spectroscopy [76]. The interest stems from the long half-life of the excited state, $t_{1/2}=6.8\times10^{-6}$ sec. Thus, the theoretical natural line width Γ_{nat} is 3.2×10^{-3} mm/sec compared to

0.97 mm/sec for ^{57}Fe. The ratio of $E\gamma/\Gamma_{nat}$ is about 30 times larger for ^{181}Ta than for ^{57}Fe. With the narrow ^{181}Ta line, the tower required in the gravitational red shift experiment with ^{57}Fe 14.4-keV gamma rays by Pound and co-workers [77,78] could be reduced to small laboratory dimensions (about 1 m). In fact, it is conceivable that instead of tracing the spectrum by a Doppler motion, the gravitational potential of source and absorber is changed. These and other interesting application possibilities, of course, have aroused the interest of many physicists from the early days in Mössbauer spectroscopy. However, the outcome of great efforts by many laboratories is rather meager. In 1964 Cohen *et al.* [79] reported a resonance effect of 0.6%. In later work by Steyert *et al.* [80] and Muir [81], the effect was increased to about 2% with a relatively broad resonance line width, hardly less than the ^{57}Fe line width. The reason for the difficulties to sharpen the 6.25-keV resonance line is most likely connected with an unresolved quadrupole interaction of the $-9/2$ excited and $+7/2$ ground state of ^{181}Ta. The lattice defects (impurities, dislocations, vacancies, etc.) introduced in the preparation of suitable sources (incorporating the parent isotope ^{181}W in a cubic lattice) and a very thin absorber (^{181}Ta is 100% abundant) in conjunction with the large quadrupole moment and change in the nuclear radius $\delta R/R$ of ^{181}Ta leads to the rather unresolved broad line. It seems that physical metallurgical considerations might find a clever way to solve the problem of a sharp ^{181}Ta resonance line. Recently, Sauer [82] reported a remarkable improvement of the ^{181}Ta resonance line and significant applications in terms of lattice defects.

3. SUMMARY

Mössbauer spectroscopy has been proven to be a valuable new tool in physical metallurgy. In selected systems experimental observations on the following topics have been reviewed: precipitation processes, phase transitions, phase identification, internal oxidation, magnetic properties, determination of Néel and Curie temperatures, near-neighbor interaction in substitutional and interstitial alloys, order–disorder properties, thin films, and superparamagnetism. One can predict that in the near future Mössbauer spectroscopy in the transmission as well as in the scattering mode will be used as a nondestructive analytical method in the investigation of various phases simultaneously present in metallic alloys. Metallurgical considerations regarding the preparation of source and absorber are discussed.

ACKNOWLEDGMENTS

This chapter was written at the Science Center, North American Rockwell Corporation, Thousand Oaks, California 91360.

REFERENCES

1. R. L. Mössbauer, Z. Physik **151**, 124 (1958).
2. R. L. Mössbauer, Naturwiss. **45**, 538 (1958).
3. A. H. Muir, Jr., K. J. Ando, and H. M. Coogan, Mössbauer Effect Data Index, 1958–1965 (Interscience, New York, 1966).
4. H. Frauenfelder, The Mössbauer Effect (W. A. Benjamin, New York, 1962).
5. M. Hansen and K. Anderko, Constitution of Binary Alloys (McGraw-Hill, New York, 1958).
6. U. Gonser, R. W. Grant, A. H. Muir, Jr., and H. Wiedersich, Acta Met. **14**, 259 (1966).
7. U. Gonser, C. J. Meechan, A. H. Muir, Jr., and H. Wiedersich, J. Appl. Phys. **34**, 2373 (1963).
8. M. Ron, A. Rosencwaig, H. Shechter, and A. Kidron, Phys. Letters **22**, 44 (1966).
9. R. M. Housley, U. Gonser, and R. W. Grant, Phys. Rev. Letters **20**, 1279 (1968).
10. A. Nagarajan and P. A. Flinn, Appl. Phys. Letters **11**, 120 (1967).
11. W. T. Chandler (Rocketdyne Division, North American Rockwell Corporation), G. Martin (Los Angeles Division, North American Rockwell Corporation), U. Gonser, R. W. Grant, R. M. Housley, A. H. Muir, Jr., and H. Wiedersich (Science Center, North American Rockwell Corporation), unpublished observations.
12. G. Shirane, D. E. Cox, and S. L. Ruby, Phys. Rev. **125**, 1158 (1962).
13. R. Bauminger, S. G. Cohen, A. Marinov, S. Ofer, and E. Segal, Phys. Rev. **122**, 1447 (1961).
14. K. Ono, Y. Ishikawa, A. Ito, and E. Hirahara, J. Phys. Soc. Japan **17**, Suppl. B-1, 125 (1962).
15. A. H. Muir, Jr. and H. Wiedersich, J. Phys. Chem. Solids **28**, 65 (1967).
16. R. J. Borg, R. Booth, and C. E. Violet, Phys. Rev. Letters **11**, 464 (1963).
17. U. Gonser, R. W. Grant, C. J. Meechan, A. H. Muir, Jr., and H. Wiedersich, J. Appl. Phys. **36**, 2124 (1965).
18. P. P. Craig and W. A. Steyert, Phys. Rev. Letters **13**, 802 (1964).
19. R. W. Grant, H. Wiedersich, and U. Gonser, Bull. Am. Phys. Soc. **10**, 708 (1965).
20. M. B. Stearns, J. Appl. Phys. **35**, 1095 (1964); ibid. **36**, 913 (1965); Phys. Rev. **147**, 439 (1966).
21. P. A. Flinn and S. L. Ruby, Phys. Rev. **124**, 34 (1961).
22. C. E. Johnson, M. S. Ridout, and T. E. Cranshaw, Proc. Phys. Soc. (London) **81**, 1079 (1963).
23. G. Shirane, C. W. Chen, P. A. Flinn, and R. Nathans, J. Appl. Phys. **34**, 1044 (1963).
24. G. K. Wertheim, V. Jaccarino, J. H. Wernick, and D. N. E. Buchanan, Phys. Rev. Letters **12**, 24 (1964).
25. K. Ohta, J. Appl. Phys. **39**, 2123 (1968).
26. H. L. Marcus, M. E. Fine, and L. H. Schwartz, J. Appl. Phys. **38**, 4750 (1967).
27. M. Cohen, Trans. AIME **224**, 638 (1962).
28. P. M. Gielen and R. Kaplow, Acta Met. **15**, 49 (1967).
29. T. Moriya, H. Ino, F. E. Fujita, and Y. Maeda, J. Phys. Soc. Japan **24**, 60 (1968).
30. J. M. Genin and P. A. Flinn, Trans. AIME **242**, 1419 (1968).
31. H. Ino, T. Moriya, F. E. Fujita, Y. Maeda, Y. Ono, and Y. Inokuti, J. Phys. Soc. Japan **25**, 88 (1968).
32. T. Shinjo, F. Itoh, H. Takaki, Y. Nakamura, and N. Shikazono, J. Phys. Soc. Japan **19**, 1252 (1964).
33. M. Ron, H. Shechter, A. A. Hirsch, and S. Niedzwiedz, Phys. Letters **20**, 481 (1966).
34. J. M. Genin and P. A. Flinn, Phys. Letters **22**, 392 (1966).
35. H. Marcus, L. H. Schwartz, and M. E. Fine, Trans. ASM **59**, 468 (1966).
36. M. Bernas, I. A. Campbell, and R. Fruchard, J. Phys. Chem. Solids **28**, 17 (1967).

37. P. M. Gielen and R. Kaplow, *Acta Met.* **15**, 49 (1967).
38. B. W. Christ and P. M. Giles, *Mössbauer Effect Methodology*, Vol. 3 (Plenum Press, New York, 1967), p. 37.
39. E. F. Makarov, V. A. Povitskii, E. B. Granovskii, and A. A. Fridman, *Phys. Status Solidi* **24**, 45 (1967).
40. M. Ron, A. Kidron, H. Shechter, and S. Niedzwiedz, *J. Appl. Phys.* **38**, 590 (1967).
41. M. Ron, H. Shechter, and S. Niedzwiedz, *J. Appl. Phys.* **39**, 265 (1968).
42. G. K. Wertheim and J. H. Wernick, *Acta Met.* **15**, 297 (1967).
43. G. P. Huffman and R. M. Fisher, *J. Appl. Phys.* **38**, 735 (1967).
44. R. Nathans, M. T. Pigott, and C. G. Shull, *J. Phys. Chem. Solids* **6**, 38 (1958).
45. K. Ono, Y. Ishikawa, and A. Ito, *J. Phys. Soc. Japan* **17**, 1747 (1962).
46. M. B. Stearns, *Phys. Rev.* **168**, 588 (1968).
47. L. Cser, J. Ostanevich, and L. Pál, *Phys. Status Solidi* **20**, 581 (1967); **20**, 591 (1967).
48. G. Shirane, C. W. Chen, P. A. Flinn, and R. Nathans, *Phys. Rev.* **131**, 183 (1963).
49. C. E. Violet and E. L. Lee, *Mössbauer Effect Methodology*, Vol. 2 (Plenum Press, New York, 1966), p. 171.
50. L. Néel, *Compt. Rend.* **228**, 604 (1949).
51. U. Gonser, H. Wiedersich, and R. W. Grant, *J. Appl. Phys.* **39**, 1004 (1968).
52. S. Nasu, T. Shinjo, Y. Nakamura, and Y. Murakami, *J. Phys. Soc. Japan* **23**, 664 (1967).
53. A. H. Muir, *Mössbauer Effect Methodology*, Vol. 4, (Plenum Press, New York, 1968), p. 75.
54. A. H. Muir, A. C. Micheletti, and M. Blander, Abstracts of the Annual Meeting of the Meteoritical Society, Moffett, Calif., 1967.
55. E. L. Sprenkel-Segel and S. S. Hanna, *Mössbauer Effect Methodology*, Vol. 2 (Plenum Press, New York, 1966), p. 113.
56. P. Imbert, *J. Phys.* **27**, 429 (1966).
57. R. M. Housley and U. Gonser, *Phys. Rev.* **171**, 480 (1968).
58. M. Blume and O. C. Kistner, *Phys. Rev.* **171**, 417 (1968).
59. R. H. Herber, *Mössbauer Effect Methodology*, Vol. 1 (Plenum Press, New York, 1965), p. 3.
60. National Bureau of Standards, Standard Reference Material 725 for Mössbauer Differential Chemical Shift for Iron-57 (see also J. J. Spijkerman, D. K. Snediker, F. C. Ruegg, and J. R. DeVoe, *NBS Misc. Publ.* 260-13).
61. G. Czjzek, J. L. C. Ford, J. C. Love, F. E. Obenshain, and H. F. Wegener, *Phys. Rev. Letters* **18**, 529 (1967).
62. G. D. Sprouse, G. M. Kalvius, and S. S. Hanna, *Phys. Rev. Letters* **18**, 1041 (1967).
63. Y. K. Lee, P. W. Keaton, Jr., E. T. Ritter, and J. C. Walker, *Phys. Rev. Letters* **14**, 957 (1965).
64. D. Seyboth, F. E. Obenshain, and G. Czjzek, *Phys. Rev. Letters* **14**, 954 (1965).
65. S. L. Ruby and R. E. Holland, *Phys. Rev. Letters* **14**, 591 (1965).
66. D. W. Hafemeister and E. B. Shera, *Phys. Rev. Letters* **14**, 593 (1965).
67. G. K. Wertheim and H. J. Guggenheim, *J. Chem. Phys.* **42**, 3873 (1965).
68. W. Trifthäuser and P. P. Craig, *Phys. Rev. Letters* **16**, 1161 (1966).
69. U. Gonser and H. Wiedersich, *J. Phys. Soc. Japan* **18**, Suppl. II, 47 (1963).
70. J. A. Stone and W. L. Pillinger, *Phys. Rev. Letters* **13**, 200 (1964).
71. J. Stephen, *Nucl. Instr. Methods* **26**, 269 (1964).
72. R. E. DeWames, T. Wolfram, and G. W. Lehman, *Phys. Rev.* **131**, 529 (1963).
73. N. E. Alekseevskii, Pham Zuy Hien, V. G. Shapiro, and V. S. Shpinel', *Zh. Eksperim. i Teor. Fiz.* **43**, 790 (1962).
74. C. J. Meechan and A. H. Muir, *Rev. Mod. Phys.* **36**, 438 (1964).
75. P. A. Flinn and S. L. Ruby, *Rev. Mod. Phys.* **36**, 352 (1964).
76. A. H. Muir and F. Boehm, *Phys. Rev.* **122**, 1564 (1961).
77. R. V. Pound and G. A. Rebka, Jr., *Phys. Rev. Letters* **4**, 337 (1960).
78. R. V. Pound and J. L. Snider, *Phys. Rev.* **140**, 8788 (1965).
79. S. G. Cohen, A. Marinov, and J. I. Budnick, *Phys. Letters* **12**, 38 (1964).

80. W. A. Steyert, R. D. Taylor, and E. K. Storms, *Phys. Rev. Letters* **14**, 739 (1965).
81. A. H. Muir, personal communication.
82. C. Sauer, *Z. Physik.* **222**, 439 (1969).

Chapter 9

Application to Biochemical Systems[1]

Leopold May

Department of Chemistry
The Catholic University of America
Washington, D.C.

Most biochemicals contain, essentially, carbon, nitrogen, and oxygen. Some include sulfur, iodine, phosphorus, and metals such as iron, calcium, and sodium. It would be very desirable to study the Mössbauer spectroscopy of the major elemental constituents, but there are conditions that limit the use of a particular nuclide. The low-lying excited state should be less than 150 keV. It is also essential that the energy of recoil be small, which is related inversely to the nuclear mass, requiring high nuclear masses. Also, the source should be readily available to the user. The first two considerations eliminate observations of the Mössbauer effect with hydrogen, carbon, nitrogen, oxygen, sulfur, and phosphorus. However, the Mössbauer spectroscopy of iodine has been applied to inorganic compounds, but not as yet to biochemically important substances such as thyroglobin.

A study has been made of bone using ^{133}Ba substituted for calcium in the bone. No ΔE_Q was found, and this was interpreted as possibly due to cancellation of the EFG by the polarized water layer or that the first state was too low [1].

The class of compounds that can be examined most readily is the metallobiochemicals. Iron, cobalt, and tin are the metals most easily studied. The latter metal is rarely found in nature, but its complexes (for example, porphyrins) can be studied since the results may be of assistance in interpreting the Mössbauer spectra of similar iron complexes. The study of cobalt-containing biochemicals is possible using ^{57}Co because it decays into ^{57}Fe. The emission spectrum is observed using an iron salt as the absorber and the

[1] Supported in part by Contract AT(30-1)-3798 with the U.S. Atomic Energy Commission.

Table 1. Biological Systems

Complete systems
    Tissue
    Bacteria
    Fluids—blood
Isolated components
    Macromolecular components
        Hemoproteins
            Respiratory—hemoglobin, myoglobin
            Electron transport—cytochrome
            Enzymes—hydroperoxidases, hydroxylases
        Nonheme proteins (NHI)
            Respiratory—hemerythrins
            Electron transport—ferredoxin
            Enzymes—dehydrogenases, oxygenases
            Iron transport—ferritin, transferrin
            Growth factor—ferrichrome
        Nucleic acids
        Polysaccharides—cellulose
    Micromolecular components
        Peptides
        Nucleotides and nucleosides
        Disaccharides
    Molecular components
        Amino acids
        Sugar
        Porphyrins

compound containing ^{57}Co as the source. The Mössbauer spectrum of vitamin B_{12} enriched with ^{57}Co has been obtained [2]. The majority of Mössbauer spectroscopic studies have been made with biochemical systems containing iron.

The study of the role of iron in biological systems can be made by examining the tissue directly, or isolated constituents of the tissue (Table 1). For example, several bacteria have been examined after being grown in cultures enriched with ^{57}Fe. It was shown that iron is involved in a nitrogen fixation by *Azotobacter vinelandii* [3]. A number of studies have been made with blood since hemoglobin, one of the major constituents of the red blood cells, is involved in the transport of oxygen and carbon dioxide. The nature of the complexing with this protein was studied by using whole blood, red blood cells, or isolated samples of this protein. Gonser *et al.* [4] studied the Mössbauer spectrum of rat red blood cells at 5°K. They found that two of the four lines in the spectrum could be found in the spectrum of oxygenated blood cells. The other two lines were identified with deoxygenated hemoglobin found in blood deoxygenated with nitrogen. Since the latter two lines were also found in blood treated with carbon dioxide, this confirmed that the carbon dioxide was not directly bound to the iron.

Studies have been made on the complexing between iron and molecular components (amino acids), micromolecular components (nucleotides), and macromolecular components such as nucleic acids, NHI proteins, and hemoproteins. Most of these studies have been with the hemoproteins, and the results of these will be used to illustrate the application of Mössbauer spectroscopy to biochemical systems.

1. HEMOPROTEINS

The structure of hemoproteins is presented diagrammatically in Figure 1. The iron is located at the center of the porphyrin ring, designated by the square. In all hemoproteins, the ligand designated as Pr is a protein chain generally considered bonded to the iron through the imidazole nitrogen of a histidine residue in a protein. In hemoglobin, the Pr ligand is globin. The ligand L can be O_2, as in oxyhemoglobin, or a protein chain, as in cytochrome c. The porphyrin ring contains four pyrrole rings (Figure 2), which are complexed to the iron atom through the pyrrole nitrogens. The porphyrin ring is essentially planar and is an aromatic system so that the four nitrogens of a pyrrole ring are considered to be equivalent. The side groups vary with different porphyrins, but protoporphyrin IX is found in hemoproteins. To obtain information concerning the bonding between the iron and the ligands in hemoproteins, it is useful to study the effect of changing the ligands on the Mössbauer spectrum.

A study of the Mössbauer spectra of the iron protoporphyrins as models for the hemoproteins yields information that is useful in the interpretation of the spectra of the various hemoproteins. Both the ferrous form, ferroprotoporphyrins (heme), and the ferri form, ferriprotoporphyrin (hematin), have been studied. The effect of the various variables, such as temperature, reveals features of the spectra that can be applied to the interpretation of the more complicated hemoproteins.

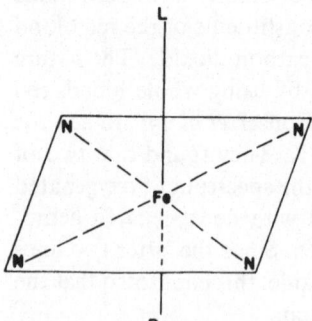

Figure 1. Diagrammatic representation of hemoprotein structure. Pr=protein, L=ligand.

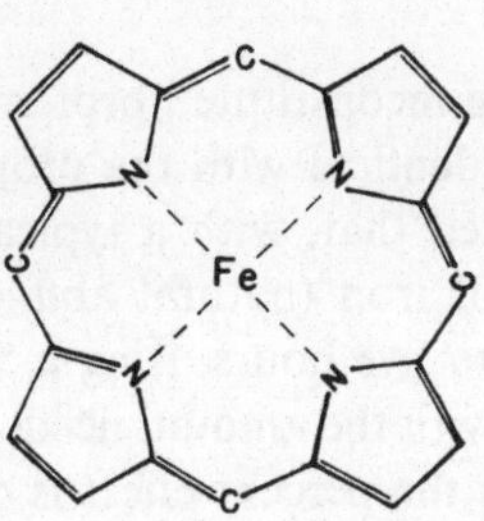

Figure 2.   Porphyrin ring structure.

2.   EXPERIMENTAL CONDITIONS

2.1.   Enrichment

The content of iron in many proteins is very small so that the observed percent effect of absorbers containing the natural amounts of ^{57}Fe is relatively small ($<1\%$) (Table 2). Improvement of the spectra can be made by increasing the content of ^{57}Fe by enriching the protein with ^{57}Fe. This permits evaluation of the Mössbauer parameters with sufficient accuracy to draw significant conclusions. If the biological system permits the use of enriched ^{57}Fe solutions as part of the fluids with which the organism is in contact, this is the preferred since the metabolism of iron is normal. For example, the hemoglobin from rat blood cells can be enriched by injecting iron citrate containing 80% ^{57}Fe [5] into the blood system of the animals. This provides an estimated improvement factor of 25 over the sample containing the natural ^{57}Fe abundance. The iron in bacteria can be enriched by including enriched iron salts in the culture medium.

For many biopolymers, this procedure is not practical, for example, beef liver catalase and horse heart myoglobin. With the latter protein, the protein part, apomyoglobin, can be removed without denaturation. Reconstitution of myoglobin can be done with apomyoglobin and enriched ^{57}Fe protohemin, yielding a reconstituted product with the same properties as the native myoglobin [6]. This is possible only with a limited number of proteins,

Table 2.   Amount of Protein Required as Absorber for Spectrum[a]

Protein	mg Fe/g protein	protein, g/10^{20} atoms Fe
Cytochrome c	4300	2.2
Myoglobin	3450	2.7
Hemoblogin	3350	,2.8
Xanthine oxidase	1400	6.7
Catalase	900	10

[a] Assuming a spectrum in twenty-four hours using a 5-mC ^{57}Co source with natural abundance of ^{57}Fe.

and the properties of the reconstituted proteins must be carefully examined to insure that they are identical with the properties of the native proteins.

It has been estimated that, with a typical spectrometer and a sample containing 10^{20} atoms of iron (natural abundance), a spectrum could be obtained in about four to five hours using a 5-mC ^{57}Co source. In Table 2 there are some proteins with the amount needed to obtain a spectrum. These calculations assume that the percent effect is constant for all these samples and approximately equal to that of inorganic iron salts. For example, a sample of 10 g of catalase would be required. However, a spectrum with 500 mg of beef liver lyophilized catalase was obtained after twenty-four hours with an effect of about 0.3 % using a 15-mC source [7].

Thus, the amount of enrichment required depends upon a number of factors: the strength of the source, instrumental factors, and the nature of the absorber. With many biochemicals, it is impracticable to enrich the iron in the sample. For example, beef liver catalase can not be labelled unless the animal (cow) is injected with radioactive iron salts requiring large and expensive amounts of radioactive material.

2.2. Effect of Thickness

Since relatively thick samples are used in the study of many bio-chemicals, the effect of thickness should be considered in the evaluation of the Mössbauer parameters. Gonser and Grant [8] have studied the thickness dependence using oxygenated blood samples. It was found that the transmission of the peak decreased and reached a minimum and the total absorption

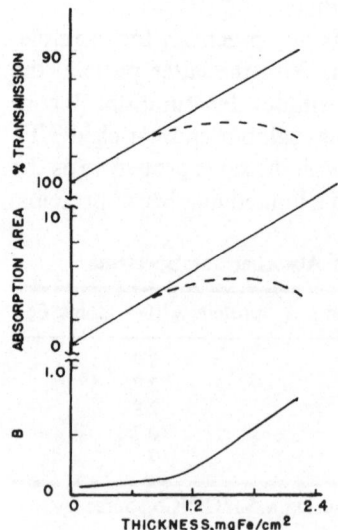

Figure 3. (a) Variation of transmission; (b) absorption area, and (c) background factor $[A_D/(A_T+A_D)]$ with absorber thickness of oxy-hemoglobin absorbers. Adapted from [8].

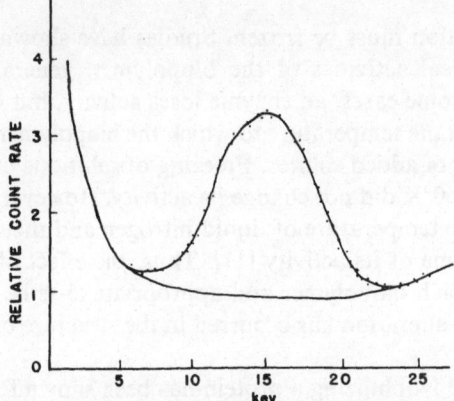

Figure 4. Definition of background A_B and $A\gamma$ from the pulse-height spectrum. Adapted from [8].

area of the two peaks in the spectrum increased to a maximum as the thickness increased (Figure 3). For these red blood cells, the resonance absorption reached a maximum at about a thickness of 1.4 cm. It is not necessarily desirable to use this thickness since the resonance photoelectric absorption is so large that long counting times may be necessary to obtain reasonable statistics unless a very strong source is used. The transmission and absorption areas can be corrected for the background determined from the pulse-height spectrum (Figure 4) by subtracting the A_B or radiation due to x rays only from the pulse-height spectrum. The background also varies with the thickness (Figure 3). These values depend upon the source, absorber, geometry, instrumentation, etc., and must be measured in each series of experiments.

2.3. State of the Absorber

Since the Mössbauer nuclei must be bound in a solid system, biochemicals can be studied only as solids or in frozen solutions. For a biopolymer, this is not its natural environment, i.e., aqueous solution at the normal temperature of the organism, which is generally near or above room temperature. It is known that drying a biopolymer at room temperature may cause conformation changes with subsequent denaturation. Even freeze-drying or lyophilization may have the same effect, but this is the safest procedure for use in preparing solid samples. The drying of deoxygenated rat hemoglobin, whose Mössbauer spectrum is a doublet, gave rise to a spectrum with three lines. These were interpreted as being due to the presence of equal amounts of both high- and low-spin ferrous ion in the dried hemoglobin [9].

In the course of preparing lyophilized samples or frozen solutions, the biopolymer solution must be frozen. Studies have shown that freezing can alter the biological activities of the biopolymer, generally with a loss of activity [10]. In some cases, an enzyme loses activity, but the loss of activity will depend upon the temperature to which the biopolymer is frozen, the pH, and the presence of added solutes. Freezing of solutions of the hemoenzyme catalase below 150°K did not change its activity. However, when the enzyme was frozen at the temperature of liquid nitrogen and then warmed to about 250°K, it lost some of its activity [11]. Thus, the effect of freezing must be examined with each biopolymer and appropriate tests be made to establish if any significant alteration has occurred in the structure of the sample being examined.

The effect of lyophilizing a protein has been shown for ferricytochrome c as well as the effect of concentration of the protein in solution [12]. The ΔE_Q of the frozen solution (0.27 g/g H_2O) was greater than that of the lyophilized sample at 150°K. As the solution was concentrated, the ΔE_Q decreased. The isomer shifts were constant, but the half-widths of lines were smaller in the solution. Very small changes were observed in the spectra of the dried and frozen solutions of the ferrocytochrome c except that the half-widths of the lines were smaller in the spectrum of the solution than in the spectrum of the powder.

The effect of low temperature on the spectrum may also be due to the usual changes found in the ΔE_Q for low-spin ferric and high-spin ferrous species, but also changes in the structure of the biopolymer such as the relative amounts of low- and high-spin forms and conformational alterations.

3.   EFFECT OF EXPERIMENTAL VARIABLES

3.1.   Effect of Temperature

As is found with iron compounds, the change in isomer shift with temperature is small and is essentially a second-order effect. The quadrupole splitting temperature variation changes with the state of the iron. For low-spin ferrous and high-spin ferric species, the temperature variation is small, whereas it is significant for the high-spin ferrous and low-spin ferric species. Using the relationships between the ΔE_Q and the EFG discussed in Chapters 4, 5, and 6, the EFG and the splittings between the d orbitals can be estimated if a structural model for the protein is assumed. As can be seen from Figure 1, a distorted octahedron may be used as a model for hemoproteins.

The asymmetry of the lines can also provide useful information, for example, the variation of the spectrum of hemin (ferriprotoporphyrin Cl) with temperature (Figure 5). The iron is in the high-spin Fe^{3+} form. As the temperature increases the pattern becomes more asymmetric with each peak

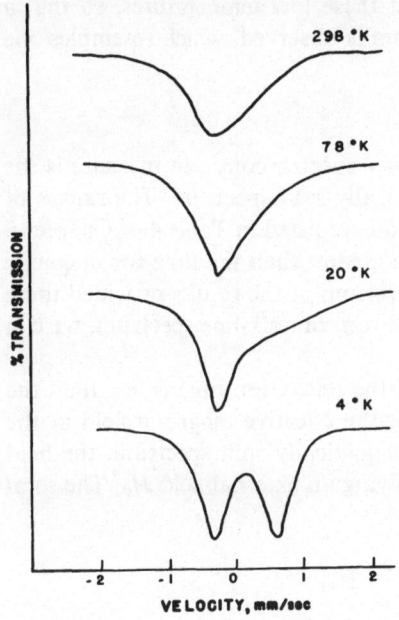

Figure 5.   Spectrum of hemin at various temperatures. Adapted from [13].

broadening and overlapping. At room temperature the spectrum consists of an asymmetric, broad band. Line broadening may be due to relaxation processes that partially average out the magnetic hyperfine interaction. The behavior of the peaks is similar to that produced by magnetic relaxation effects except that the lines broaden as the temperature decreases. With hemin, the sharpening of the lines increases as the temperature decreases. Blume [13] showed that this is due to spin–spin relaxation together with the thermal excitation of higher Kramers doublets. At the low temperatures, the ions are in the ground state ($S= \pm1/2$). With increasing temperature the excited states ($S= \pm3/2, \pm5/2$) with slower relaxation rates are populated. Hence, the nuclei whose ions are in these excited states produce asymmetric spectra. In addition, the dominant relaxation mechanism, spin–spin relaxation time, becomes slower as the temperature is increased. The excited states have larger interval magnetic fields, which tend to broaden the lines.

Another relaxation process, spin-lattice relaxation, may be responsible for the line broadening as mentioned above with the broadening of the lines increasing as the temperature decreases. If the relaxation time is sufficiently long compared to the characteristic time for magnetic hyperfine interactions, the typical six-line spectrum may be observed (see Chapter 5). With many paramagnetic iron-containing biopolymers, the spectrum may contain six lines at liquid-helium temperatures or below. For many proteins, the relaxation time is too short even at these low temperatures, so that a characteristic hyperfine splitting spectrum is observed, which resembles one large broadened line.

3.2. Effect of Magnetic Field

One of the quantities that Mössbauer spectroscopy can measure is the internal magnetic field from the magnetically split spectrum. The ranges of values for the various charge states of iron are listed in Table 4 of Chapter 6 (p. 130). If the spin relaxation times are greater than the time for magnetic hyperfine interactions, then a six-line spectrum of the results provided there is a measurable internal magnetic field. From this six-line spectrum, we can determine the strength of the field.

If this condition is not present and the relaxation time is less than the time for magnetic hyperfine interaction, the effective magnetic field at the nucleus will be zero. To produce the magnetically split spectrum, the field at the nucleus must be increased by applying an external field H_a. The total field H at the nucleus then is

$$H_a + \frac{<S>}{S} H_n,$$

where $<S>/S$ is the magnetization of the electron spins and H_n is the

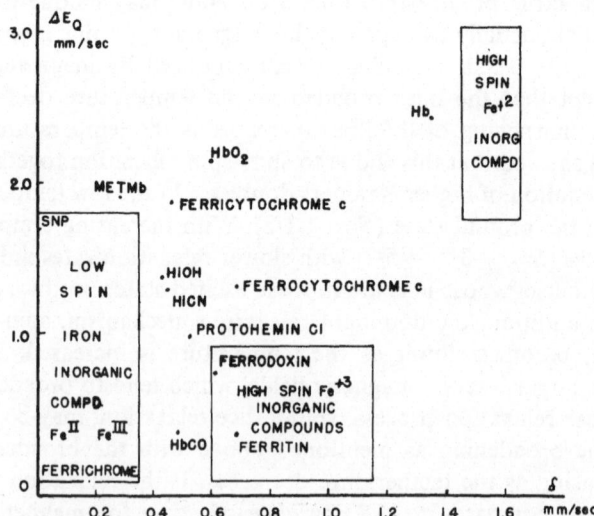

Figure 6. Scatter plot of iron biochemicals and inorganic compounds. Correlation of quadrupole splitting ΔE_Q with isomer shift δ. The isomer shift is relative to sodium nitroprusside. Adapted from [18].

internal magnetic field. At low values of magnetization, asymmetrical line broadening was observed, but at high values of magnetization, well resolved spectra were observed. For example, Johnson [14] was able to measure the internal magnetic field and the EFG from measurements on hemin.

The isomer shifts for both the low-spin forms of ferrous and ferric species are in the same range so that it is difficult to distinguish the charge state of the iron from this parameter alone (Figure 6). The differences in the ranges of ΔE_Q are very small for these biochemicals. The low-spin ferriproteins ($S=1/2$) may show hyperfine splitting at low temperature, but the low-spin ferroproteins will not since they are diamagnetic. The application of an external magnetic field will also cause hyperfine splitting in the spectrum of a low-spin ferriprotein but have no effect upon the spectrum of a diamagnetic low-spin ferroprotein.

4. SPIN–SPIN EQUILIBRIUM

As can be seen from the scatter plot in Figure 6, most of the hemoproteins appear in regions outside of those for the iron inorganic compounds and the NHI proteins. This suggests that the iron may exist in intermediate spin states in these hemoproteins as suggested by Erickson [15] from Mössbauer spectral results for iron compounds. The measurements of the optical spectra and magnetic susceptibilities of a number of low-spin ferrihemoproteins suggests that the iron site can be in the intermediate spin state ($S=3/2$) between the low-spin ($S=1/2$) and high-spin ($S=5/2$) states or that an equilibrium exists between both states. George et al. [16] proposed that the latter was present from these measurements. They used the results of the fluoride complexes as the values for the high-spin species, and values for the low-spin species were taken from measurements of the cyanide complexes. Both of these measurements (spectral and magnetic susceptibility) on the native proteins give only single values for a multispin system. Electron spin resonance and Mössbauer spectral measurements can detect the two different species, but ESR measurements can be very inaccurate in the determination of the relative concentrations of both species.

Lang et al. [17] measured the Mössbauer spectra of protoheme and mesoheme ferricytochrome c peroxidases at 4.2 and 195°K. From the parameters extracted from the spectra, they were able to identify both high- and low-spin species and measure the relative concentrations of both forms of the enzyme. For the protoheme enzyme, the high-spin fraction increased from 0.25 at 4.2°K to 0.63 at 195°K. The lines due to the two-spin species were easily discernable at 195°K, but not at the lower temperature. Application of an external magnetic field was necessary at the lower temperature · to determine the relative concentrations of the two-spin species.

REFERENCES

1. J. H. Marshall, *Phys. Med. Biol.* **13**, 15 (1968).
2. A. Nath, M. Harpold, and M. P. Klein, *Chem. Phys. Letters* **2**, 471 (1968).
3. G. V. Novikov, L. A. Syrtsova, G. I. Likhtenshtein, V. A. Trukhtanov, V. F. Rachek, and V. I. Gol'danskii, *Dokl. Akad. Nauk. SSR* **181**, 1170 (1968) (Russian); *Proc. Acad. Sci. USSR, Phys. Chem. Sect.* **181**, 590 (1968) (English).
4. U. Gonser, R. W. Grant, and J. Kregzde, *Science* **143**, 680 (1964).
5. G. Lang and W. Marshall, *Proc. Phys. Soc.* **87**, 3 (1966).
6. W. S. Caughey, W. Y. Fujimoto, A. J. Bearden, and T. H. Moss, *Biochemistry* **5**, 1255 (1966).
7. L. May and Geraldine M. Hasco, Abstr. No. 124, 156 National Meeting, American Chemical Society, 1968.
8. U. Gonser and R. W. Grant, *Biophys. J.* **5**, 823 (1965).
9. R. W. Grant, J. A. Cape, U. Gonser, L. E. Topol, and P. Saltman, *Biophys. J.* **6**, 651 (1967).
10. *Federation Proc.* **24**, Suppl. No. 15 (1965).
11. R. B. Pennell, *Federation Proc.* **24**, Suppl. No. 15, S-269 (1965).
12. R. Cooke and P. Debrunner, *J. Chem. Phys.* **48**, 4532 (1968).
13. M. Blume, *Phys. Rev. Letters* **18**, 305 (1967).
14. C. E. Johnson, *Phys. Letters* **21**, 491 (1966).
15. N. E. Erickson, *Advan. Chem. Ser.* **68**, 86 (1967).
16. P. George, J. Beetlestone, and J. S. Griffith, in *Haematin Enzymes*, J. E. Falk, R. Lemberg and R. K. Morton, Eds. (Pergamon, Oxford, 1961), p. 105; *Rev. Mod. Phys.* **36**, 441 (1964).
17. G. Lang, T. Asakura, and T. Yonetani, *J. Phys. C* **2**, 2246 (1969).
18. L. May, *Advan. Chem. Ser.* **68**, 52 (1967).

Appendix A

Nomenclature of Mössbauer Spectroscopy

This Appendix includes the nomenclature used in the text. Other nomenclature compilations can be found in NBS Misc. Publ. 260-13 (1967) and in the report of the Mössbauer Spectroscopy Task Group, ASTM Committee E-4, R. H. Herber, Chairman, 1969.

Name	Symbol	Units	Definition
Isomer shift	δ	mm/s	Displacement of the center of resonance spectrum from reference point
Quadrupole splitting	ΔE_Q	mm/s mc	Hyperfine interaction (line splitting) between the nuclear quadrupole moment and the electric field gradient
Line width	Γ Γ_{exp}	mm/s	The full width at half maximum of the experimental Mössbauer line
Natural line width	Γ_{nat}	mm/s	Is h/τ, where h is $\hbar$, Planck's constant divided by 2π, and τ is the mean lifetime of the excited state; $=4.55 \times 10^{-16}/t_{1/2}$ where $t_{1/2}$ is the half lifetime of the excited state
Extrapolated line width	Γ_0	mm/s	The line width (Γ) extrapolated to zero thickness
Magnitude of the effect	ε	percent	$\dfrac{[I(\infty) - I(0)]}{I(\infty)} \times 100$, where $I(\infty)$ is the count rate at which resonance effect is negligible and $I(0)$ is the count rate at resonance maximum
Recoil-free fraction	f	percent	Fraction of all gamma rays of the Mössbauer transition that are emitted (f_s) or absorbed (f_a) without recoil
Mössbauer thickness	T		Effective thickness of a source (T_s) or absorber (T_a); $T_a = f_a \sigma_0 \, na$, where

191

σ_0 is the resonance cross section in cm^2, n is the number of all Mössbauer atoms, and a is the fractional abundance of Mössbauer isotope

Resonance cross section	σ_0	cm^2	The cross section for resonance absorption
Internal magnetic field	H_n	oersted	Value of field at the nucleus
External magnetic field	H_e	oersted	Value of applied field
Energy of the gamma radiation	E_r	keV	Mean energy of the gamma radiation
Electric field gradient tensor	EFG		Tensor describing the electric field gradient specified by η and V_{zz} in addition to the Euler angles specifying the tensor orientation
Principal component the electric field gradient tensor	V_{zz}	V/cm^2	$V_{zz} = \partial^2 V / \partial z^2 = eq$
Asymmetry parameter	η		$(V_{xx} - V_{xy})/V_{zz}$
Nuclear quadrupole moment	Q	barn(b)	Parameter describing the shape of the nuclear charge distribution

Appendix B

Bibliographic Sources

Mössbauer Effect Data Index 1958–1965, A. H. Muir, Jr., K. J. Ando, and Helen M. Coogan (Interscience, New York, 1966) 351 pp. + xviii.
This listing of references is essential to all serious workers in this field. For each isotope, there is included a summary of nuclear and Mössbauer properties. In addition, one listing of the references concerned with this isotope is made according to the source used and another according to the absorber used. In additional sections, the references are listed according to subject matter, for example, analysis, instrumentation, etc. An author index is also included and a chronological listing of the references, each of which is given a code consisting of the last two digits of the year of publication, a letter corresponding to the initial of the last name of the senior author, and two numbers that are arbitrary. It is planned to add additional volumes to keep the references current.

Bibliography of Papers on Recoiless Radiation, Rev. Mod. Phys. **36**, 472–503 (1964).
A listing of papers through August 1963.

The Mössbauer Effect, Bibliographical Series No. 16, International Atomic Energy Agency, Vienna, 1965, 137 pp. + xvi.
This bibliography includes a brief abstract of each paper. The papers are listed according to subjects such as theory, application to nuclear physics, etc. Cross-indexing is also provided. An author index is included. It contains 776 references published from March 1958 to March 1964.

Mössbauer Spectrometry, J. R. Devoe and J. J. Spijkerman, *Anal. Chem.* **38**, 382R–393R (1966); **40**, 472R–489R (1968); **42**, 366R–388R (1970).
These are the first three issues of a continuing biennial review series. It compiles references for the interval between the publication of the reviews. The references are collated for each isotope with a description of subject or material studied in a table.

Index of Publications in Mössbauer Spectroscopy of Biological Materials, L.

May, Department of Chemistry, The Catholic University of America, Washington, D. C. 20017.

The index consists of a peek-a-boo card system for retrieval of this information. The system permits you to search for the information such as the Mössbauer parameters obtained with various biochemicals. Included is a listing of the various references used in preparing the punch cards. The listing of the references includes the authors, the title of the article, bibliographical data, and the *Mössbauer Effect Data Index* number. Issued every six months.

Appendix C

Selected References on Mössbauer Spectroscopy

BOOKS

A. Abragam, *L'éffect Mössbauer et ses applications a l'etude des champs Internes* (Gordon and Breach, New York, 1964).

J. Danon, *Lectures on The Mössbauer Effect* (Gordon and Breach, New York, 1968).

H. Frauenfelder, *The Mössbauer Effect* (W. A. Benjamin, New York, 1962).

V. I. Gol'danskii, *The Mössbauer Effect and Its Applications in Chemistry*, (Consultants Bureau, New York, 1964; Van Nostrand, 1966).

V. I. Gol'danskii and R. H. Herber, Eds., *Chemical Applications of Mössbauer Spectroscopy* (Academic Press, New York, 1968).

H. Wegener, *Der Mössbauer-Effekt und seine Anwendung in Physik und Chemie* (Bibliographisches Institut, Mannheim, 1965).

G. K. Wertheim, *Mössbauer Effect: Principles and Applications* (Academic Press, New York, 1964).

V. S. Shpinel', Gamma-Ray Resonance in Crystals (Izd. "Nauka," Moscow, 1969). [В. С. Шпинель, Резонанс Гамма-лучей в Кристаллах (Издательство «Наука», Москва, 1969).]

REVIEW ARTICLES

Catalysis

W. N. Delgass and M. Boudart, *Catal. Rev.* **2,** 129 (1968).

Chemistry

P. R. Brady, P. R. F. Wigley, and J. F. Duncan, *Rev. Pure Appl. Chem. (Australia)* **12,** 165 (1962).

E. Fluck, W. Kerler, and W. Neuwirth, *Angew. Chem., Intern. Ed.* **2,** 277 (1963).

E. Fluck, *Adv. Inorg. Chem. Radiochem.* **6,** 433 (1964).

J. F. Duncan and R. M. Golding, *Quart. Rev.* **19,** 36 (1965).

R. H. Herber, *J. Chem. Educ.* **42,** 180 (1965).
E. Fluck, *Fortschr. Chem. Forsch.* **5,** 399 (1966) (German).
J. R. Devoe and J. J. Spijkerman, *Anal. Chem.* **38,** 382R (1966); **40,** 472R (1968); **42,** 366R (1970).
R. H. Herber, *Ann. Rev. Phys. Chem.* **17,** 261 (1966).
N. N. Greenwood, *Chem. Britain* **3,** 56 (1967).
R. H. Herber, *Prog. Inorg. Chem.* **8,** 1 (1967).
J. J. Spijkerman, *Tech. Inorg. Chem.* **7,** 71 (1968).
D. A. Shirley, *Ann. Rev. Phys. Chem.* **20,** 25 (1969).

Metallurgy

U. Gonser, *Mat. Sci. Eng.* **3,** 1 (1968).

Physics

R. L. Mössbauer, *Ann. Rev. Nucl. Sci.* **12,** 123 (1962).
A. J. F. Boyle and H. E. Hall, *Rept. Progr. Phys.* **25,** 441 (1962).
S. de Benedetti, F. deS. Barros, and G. R. Hay, *Ann. Rev. Nucl. Sci.* **16,** 31 (1966).
Mössbauer Effect, Selected Reprints (American Institute of Physics, New York, 1963).

SYMPOSIA

D. M. J. Compton and A. H. Schoen, Eds., *Transactions of the Second Conference on the Mössbauer Effect* (John Wiley and Sons, New York, 1962).
Proceedings of the Dubna Conference on the Mössbauer Effect, **1962** (Consultants Bureau, New York, 1963).
A. J. Bearden, Ed., *Third International Conference on the Mössbauer Effect,* **1963,** in *Rev. Mod. Phys.* **36,** 333 (1964).
I. J. Gruverman, Ed., *Mössbauer Effect Methodology* (Plenum Press, New York), Vols. 1–5, 1966–70.
Applications of the Mössbauer Effect in Chemistry and Solid-State Physics, Intern. Atomic Energy Agency, Tech. Rept. Ser. No. 50, Vienna, 1966.
The Mössbauer Effect, Symp. Faraday Soc. **No. 1,** 1967.
The Mössbauer Effect and Its Applications in Chemistry, Advan. Chem. Ser. **68** (1967).

Index

A

Absorber, biochemicals, 185–186
Absorbers, materials for, 33
Absorber thickness, 33, 66, 184
    and intensity, 66
    and line broadening, 33, 66
    in blood cells, 184
Absorption and SnO_2 concentration,
    34
Accuracy of parameters, 28
Anharmonicity, 118
Asymmetry parameter, 57, 60, 78,
    133
Attenuation   of   non-Mössbauer
    radiation, 31–32
    filters for (table), 31
Attenuation coefficients for 14.4 and
    23.8 keV γ-rays, 32

B

^{133}Ba in bone, 180
Backscattering
    geometry, 39
    in metallurgy, 161–163
    of iron films, 40–41
    penetration depth, 40
    proportional counter for, 38–39
    spectrum of iron films, 41
    spectrum of 302 stainless steel,
    162
$BaSnO_3$, standard for tin, 143
Biochemicals, ^{57}Fe
    internal magnetic field, 188
    isomer shift, 188
    quadrupole splitting, 188

Blood cells, 181
    effect of thickness on spectrum,
    184–185
Bone, ^{133}Ba in, 180

C

$[\pi\text{-}C_5H_5Fe(CO)_2]_2$ $SnCl_2$ and related
    compounds, 150–152
Calibration, 29–30
    by Michelson interferometer,
    29
    by Moiré technique, 30
    for iron
        with α-Fe, 173
        with sodium nitroprusside,
        139, 173
    for tin, with SnO_2, 140, 143
Clebsch–Gordon coefficients, 59
^{57}Co
    energy spectrum, 24
    sources, 174–175
    spectrum in diamond, 73
Compton scattering, 33
Computation of spectra, 41–44
    by constraint least-square analy-
    sis, 43
    by least-square analysis, 42
    by stripping technique, 173
    from theoretical model, 42
Computer stripping of spectra, 173
Conformational studies, 147–152
Cosine smearing, 28
Coulomb excitation–recoil implanta-
    tion method, 72

197